International Federation of Automatic Control

SAFETY OF COMPUTER CONTROL SYSTEMS

IFAC Conference Proceedings

ATHERTON: Multivariable Technological Systems
BANKS & PRITCHARD: Control of Distributed Parameter Systems
CICHOCKI & STRASZAK: Systems Analysis Applications to Complex Programs
CRONHJORT: Real Time Programming 1978
CUENOD: Computer Aided Design of Control Systems
DE GIORGIO & ROVEDA: Criteria for Selecting Appropriate Technologies under Different Cultural, Technical and Social Conditions
DUBUISSON: Information and Systems
GHONAIMY: Systems Approach for Development
HASEGAWA & INOUE: Urban, Regional and National Planning - Environmental Aspects
LEONHARD: Control in Power Electronics and Electrical Drives
MUNDAY: Automatic Control in Space
NIEMI: A Link Between Science and Applications of Automatic Control
NOVAK: Software for Computer Control
OSHIMA: Information Control Problems in Manufacturing Technology
RIJNSDORP: Case Studies in Automation related to Humanization of Work
SAWARAGI & AKASHI: Environmental Systems Planning, Design and Control
SINGH & TITLI: Control and Management of Integrated Industrial Complexes
SMEDEMA: Real Time Programming 1977

NOTICE TO READERS

Dear Reader

If your library is not already a standing order customer or subscriber to this series, may we recommend that you place a standing or subscription order to receive immediately upon publication all new volumes published in this valuable series. Should you find that these volumes no longer serve your needs your order can be cancelled at any time without notice.

ROBERT MAXWELL
Publisher at Pergamon Press

SAFETY OF COMPUTER CONTROL SYSTEMS

Proceedings of the IFAC Workshop, Stuttgart, Federal Republic of Germany, 16-18 May 1979

Edited by

R. LAUBER

Institute for Control Engineering and Process Automation of the University of Stuttgart, Federal Republic of Germany

Published for the

INTERNATIONAL FEDERATION OF AUTOMATIC CONTROL

by

PERGAMON PRESS

OXFORD · NEW YORK · TORONTO · SYDNEY · PARIS · FRANKFURT

U.K.	Pergamon Press Ltd., Headington Hill Hall, Oxford OX3 0BW, England
U.S.A.	Pergamon Press Inc., Maxwell House, Fairview Park, Elmsford, New York 10523, U.S.A.
CANADA	Pergamon of Canada, Suite 104, 150 Consumers Road, Willowdale, Ontario M2J 1P9, Canada
AUSTRALIA	Pergamon Press (Aust.) Pty. Ltd., P.O. Box 544, Potts Point, N.S.W. 2011, Australia
FRANCE	Pergamon Press SARL, 24 rue des Ecoles, 75240 Paris, Cedex 05, France
FEDERAL REPUBLIC OF GERMANY	Pergamon Press GmbH, 6242 Kronberg-Taunus, Pferdstrasse 1, Federal Republic of Germany

Copyright © IFAC 1980

All Rights Reserved. No part of this publication may be reproduced, stored in a retrieval system or transmitted in any form or by any means: electronic, electrostatic, magnetic tape, mechanical, photocopying, recording or otherwise, without permission in writing from the copyright holders.

First edition 1980

British Library Cataloguing in Publication Data
Safety of computer control systems
1. Automatic control - Data processing - Congresses
I. Lauber, R II. International Federation of Automatic Control
629.8'95   TJ212   80-40183
ISBN 0-08-024453-X

These proceedings were reproduced by means of the photo-offset process using the manuscripts supplied by the authors of the different papers. The manuscripts have been typed using different typewriters and typefaces. The lay-out, figures and tables of some papers did not agree completely with the standard requirements; consequently the reproduction does not display complete uniformity. To ensure rapid publication this discrepancy could not be changed; nor could the English be checked completely. Therefore, the readers are asked to excuse any deficiencies of this publication which may be due to the above mentioned reasons.

The Editor

Printed in Great Britain by A. Wheaton & Co., Ltd., Exeter

IFAC WORKSHOP ON SAFETY OF COMPUTER CONTROL SYSTEMS

Organized by
VDI/VDE Gesellschaft für Mess- und Regelungstechnik (GMR)

Sponsored by
The International Federation of Automatic Control

In co-operation with
EWICS—European Workshop on Industrial Computer Systems (Purdue Europe)
CSNI—Committee for Safety of Nuclear Installations of OECD Nuclear Energy Agency

International Program Committee
R. Lauber, F.R.G. (Chairman)
A. Aitken, U.K.
D. R. Bristol, U.S.A.
W. D. Ehrenberger, F.R.G.
H. Frey, Switzerland
R. Genser, Austria
A. A. Levene, U.K.
L. Motus, U.S.S.R.
M.-F. Pau, France
M. Paul, Austria
J. R. Taylor, Denmark
R. W. Yunker, U.S.A.

CONTENTS

Preface ix

Session 1. Opening of the Workshop

Introduction into the subject of the workshop 1
R. Lauber

Safety and reliability — their terms and models of complex systems 3
H. H. Frey

Experience with computers on some UK power plants 11
B. K. Daniels, A. Aitken and I. C. Smith

Session 2. Project Management and Documentation

Guidelines for the documentation of safety related computer systems 33
A. A. Levene

Safety considerations in project management of computerized automation systems 41
H. Trauboth and H. Frey

Standards for the production of high quality systems 51
G. P. Mullery

Session 3. Systems Design and Interfaces

Functional redundancy to achieve high reliability 59
M. H. Gilbert and W. J. Quirk

Communication protocols for the PDV bus in network representation 65
T. Grams and M. Schäfer

Session 4. Software Diversity

Software diversity in reactor protection systems: An experiment 75
L. Gmeiner and U. Voges

On a diversified parallel microcomputer system 81
R. Konakovsky

An investigation of methods for production and verification of highly reliable software 89
G. Dahll and J. Lahti

Session 5. Software Testing

A survey of methods for the validation of safety related software 95
U. Voges and J. R. Taylor

An experience in design and validation of software for a reactor protection system 103
S. Bologna, E. de Agostino, A. Mattucci, P. Monaci and M. G. Putignani

Graphs of data flow dependencies 117
P. Puhr-Westerheide

Safety program validation by means of control checking — 129
W. Ehrenberger and S. Bologna

Session 6. Safety in Digital Control

A process computer for experimental use — 139
H. Holt

Control of nuclear reaction by pattern recognition methods — 149
B. Dubuisson and P. Lavison

Session 7. System Specification

Specification, design and implementation of computer-based reactor safety systems — 153
M. J. Cooper, W. J. Quirk and D. W. Clark

Experience with a specification language in the dual development of safety system software — 161
H. So, C. Nam, H. Reeves, T. Albert, E. Straker, S. Saib and A. B. Long

Session 8. Hardware Design and Testing I

Overview of hardware-related safety problems of computer control systems — 169
W. Schwier

The combined role of redundancy and test programs in improving fault tolerance and failure detection — 179
H. Schüller and C. Santucci

A fail-safe comparator for analogous signals within computer control systems — 187
G. H. Schildt

Session 9. Hardware Design and Testing II

Failure detection in microcomputer systems — 195
B. Courtois

Test policy vs. maintenance policy and system reliability — 201
L. F. Pau

Optimisation of a servosystem — 207
T. Adjarov

Session 10. Systems Approval and Licencing

Inspection of process computers for nuclear power plants — 213
G. Glöe

Author Index — 219

PREFACE

The idea for this workshop was born in the technical committee TC 7 "Safety and Security" of the European Workshop on Industrial Computer Systems (known as Purdue Europe). Therefore, it is not surprising that the majority of the papers are closely related to the work of this committee. Also proposals of CSNI were incorporated in the topics.

The workshop was the first of its kind to discuss, on an international scale, the problems inherent in the application of computer control (hardware and software) to such automated systems where hazards to human life, well-being, the environment, and property must be prevented.

There were 101 participants present from 15 different countries. The topics were covered in 10 technical sessions. In addition to these sessions, there was an informal meeting in the city hall (by invitation of the mayor of Stuttgart) on the evening of the first day, and a presentation and demonstration of the specification and design technique EPOS on the second evening (by invitation of the Institute for Control Engineering and Process Automation of the University of Stuttgart). On the afternoon of the last day, there was a very interesting visit to the computer controlled operating center of the local transport railway system (S-Bahn) by invitation of the "Deutsche Bundesbahn" in Stuttgart.

Though the main application of the contributions were in the fields of nuclear reactors and of traffic systems (i.e. railway systems), the methods discussed were applicable to all kinds of safety-related systems. This indicates that there is emerging a new application-independent discipline of safe computing. The question of main concern on the conference was the validation and licensing of safety-related software. It seems that the problem of finding adequate strategies on how to design, to build and to validate the safety of software systems will continue to be a challenge for further scientific work, and that - hopefully - progress in this field may be presented and discussed in a future SAFECOMP Workshop.

I appreciate the good cooperation with the "VDI/VDE Gesellschaft fuer Meß- und Regelungstechnik" and especially with Mr. M.A. Kaaz. Also, I owe thanks to all members of the Institute for Control Engineering and Process Automation for their assistance in the organization of the workshop and in the preparation of these proceedings.

R. Lauber

INTRODUCTION INTO THE SUBJECT OF THE WORKSHOP

R. Lauber

*Institute for Control Engineering and Process Automation,
University of Stuttgart, Federal Republic of Germany*

PROBLEM STATEMENT

When computer control is applied to safety-related technical processes like, for example,

- railway systems
- aircraft landing systems
- nuclear power stations
- chemical reactors
- elevators
- cranes

etc., faults in the computer system may cause hazards to human life. Therefore, in many countries, a certification is required, based on an assessment by a licensing authority. Strategies must be found to design and build control computer systems in such a way that a formal approval and certification is possible.

CLASSES OF OBSTACLES TO SAFE COMPUTING

In control computer systems, safety problems may arise from failures and errors of many types (fig. 1).

Fig. 1: Obstacles to safe operation of control computer systems

Considering the causes of possible hardware or software faults, three different categories may be distinguished (fig. 2):

- hardware faults due to physical or chemical failure mechanisms

- hardware design errors and software errors due to "human failure mechanisms" (mistakes, oversights, misunderstandings, wrong interpretations etc. committed by humans)

- software faults due to sporadic hardware failures. This effect is known as "software poisoning".

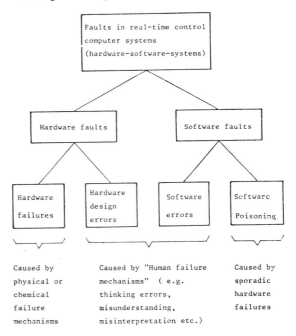

Fig. 2: Categories of causes of faults in real-time computer systems

The errors induced by "human failure mechanisms" may be further classified according to the design level into

- system requirement specification errors
- system design errors
- functional design errors
- hardware or software preliminary design errors

- coding errors.

STRATEGIES FOR SAFETY-RELATED COMPUTER SYSTEMS

There are two basic strategies to avoid possible dangers induced by computer hardware or software faults (fig. 3):

-1 to eliminate the failure mechanisms (physical, chemical or human failure mechanisms) and to aim at a perfect system without faults.
Therefore, this approach is often called the "perfectionist approach".

-2 to use a non-perfect system and to tolerate failure mechanisms, but to apply special means in order to prevent the failure mechanisms from causing any danger. This approach is called the "fault-tolerant approach".

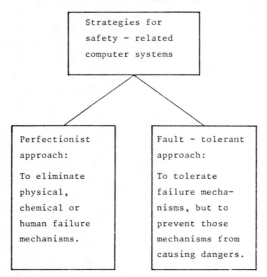

Fig. 3: Basic strategies for safety-related computer systems (hardware and software)

Because of the complexity of computer hardware and software, the favorite strategy in most applications is the fault-tolerant approach, realized by

- redundant hardware systems
- divers software systems.

But to compare redundant hardware or divers software channels, common non-redundant input and output parts are necessary, which require to be designed using methods of the "perfectionist approach", such as, in the case of software, formal specification, verification and correctness proof techniques, systematic and statistical test methods, path analysis methods etc.

Thus, in many real systems, both the strategies of the perfectionist and of the fault-tolerant approach have to be combined.

CONCLUSION

In control computer systems, errors and failures may cause dangers to human life. The papers to be presented to this workshop will show techniques and strategies to prevent those dangers and to form a basis for systems acceptance, approval, and licensing.

REFERENCES

[LBR 75] Lauber, R.: Safe Software by Functional Diversity. Purdue Europe TC Safety and Security, Paper No. 37 (Oct. 1975)

[TAY 76] Taylor, J.R.: Strategies for improving Software System reliability. Purdue Europe TC Safety and Security, Paper No. 48 (Oct. 1976)

[ENG 76] Engler, F.: Safe Software Strategies. Purdue Europe TC Safety and Security, Paper No. 89 (Oct. 1976)

[EHR 76] Ehrenberger, W.D.: Aspects of Software Verification: Statistical methods and Analytical methods. Purdue Europe TC Safety and Security, Paper No. 53 (March 1976)

[LBR 78] Lauber, R.: Software Strategies for Safety-related computer systems, Purdue Workshop, Purdue University Lafayette, Ind. USA (Sept. 1978)

SAFETY AND RELIABILITY — THEIR TERMS AND MODELS OF COMPLEX SYSTEMS

H. H. Frey

Quality Assurance Department, Div. E, Brown Boveri & Co. Ltd., CH-5401 Baden, Switzerland

Abstract. Some terms concerning reliable and safe industrial computer systems are qualitatively defined (European Purdue Workshop, TC Safety and Security). The fundamental model of reliability and safety of complex systems is defined by the set of system failure modes whereby the basic difference between hardware and software failure behaviour is discussed. Furthermore the characteristics and application of the model are illustrated and discussed by the example of a car in the traffic, whereby utility versus cost is evaluated by the term "system effectivness", the comprehensive figure of merit of complex systems.

Keywords. System safety, system reliability, reliability model, safety model, terms and definitions of, hardware failures, software failures, systemeffectiveness.

INTRODUCTION

In order to deal with a new field in science or technique, it is necessary to build up a vocabulary of special terms to allow efficient, unequivocal communication. This was also the case for the field of computer and computer systems.

But, engineers dealing with the new field of reliable and safe industrial computer systems find themselves still in a rather awkward position of trying to make themselves clear in specifying, engineering and verifying these complex systems.

This contribution will try to shed some more light on models, terms and definition, related to the safety and reliability of industrial Computer Systems or complex systems. A "Glossary of Terms and Definitions" evolving from the intensive discussions and work of the "European Purdue Workshop" TC "Safety and Security" and its Sub-Group "Terms and Definitions" is presented in [12].

The qualitative definitions of the most important terms in the context of this contribution are [12]*:

---

*Acknowledgement: At this point I also like to thank all my colleagues in the European Purdue Workshop TC Safety and Security for all their work and support on Terms and Definitions.

Safety:	Avoidance of or protection against, danger to life and limb, the environment and property arising from failures.
Safe:	Protection against all classes of threat, whether accidental or deliberate to an installation or to the plant which it controls.
Security:	Protection against all classes of threat, whether accidental or deliberate, to an installation or to the plant which it controls.
Secure:	Adequately protected against all classes of threat.
Reliability:	The ability of an item to perform a required function under stated conditions for a stated period of time.
Availability:	The ability of an item to perform its intended function when required.
Correct:	In accordance with the intended specification.
Error:	a) A human mistake or omission. b) The difference between the actual and the desired output of control systems or computer.
Failure:	The termination of an item to perform its specified function.
Failure mode:	The way an item or system fails.

Fault tolerance: The ability of a system to perform correctly in the presence of a limited number of faults.

Verification: Demonstrating the truth or correctness of a statement system, etc., by special investigation or comparison of data.

THE BASIC MODEL OF SAFE AND RELIABLE SYSTEM PERFORMANCE

Mathematical model

The basic assumption is that the functioning of the system deterministically depends on the correct function or correctness of its components or elements (hardware, software, human factors, data, etc.). An obvious way to present the functional relationship between the system components and system operation is by a state space.

Let the "System" state space Z be defined by the set of all observable system states X_j, j=1,.....,m, this set being partitioned into at least two sub-sets S (functional, reliable, safe states, etc.) and F (failure modes or unreliable, unsafe states, etc.); thus

$$S \cup F = Z \quad (1)$$
$$S \cap F = \emptyset \quad \text{(disjoint)} \quad (2)$$

Similar to the system a state space is accordingly defined for each "subsystem" and "component" and also partitioned into at least two sub-sets, thus describing the functional hierarchy of the system. (Mathematical formulation see [1] - [3],[11].)

Furthermore, the system's environment or the process it controls, must be described. That is, the "Environment" or "Process" is defined by a set of events, operating modes and their relationship to the "System". The relationship is given by the sequence of states in time of the System, which will exert a particular effect on the proper (safe, reliable, available, etc.) operation of the environment or the Process. On the other hand, the operating mode and certain events occurring independently of the system in the Environment may disturb the correct performance of the System and be the cause of failures.

Thus, only knowledge of the system states or the System failures modes, their effects on the correct operation of the Environment and the environmental factors or events disturbing the correct system performance, will be sufficient to define qualitatively the safety and reliability of the System in question.

For example, the "System" may be a safeguard in a nuclear power plant, where power plant operation is regarded as the "Environment" or "Process", or e.g. [4]. This abstract formulation will be elaborated upon by an example in the next section.

THE SYSTEM "AUTOMOBILE" AND ITS ENVIRONMENT, AN EXAMPLE

Thinking of a car (=System), one may associate with it many terms concerning its environment and life-cycle. These are, e.g.:

Table I: Terms associated with a car

- speed
- gasoline
- utility
- cost
- performance
- safety
- power
- failure

- crash
- accident
- pedestrians
- traffic
- traffic lights
- freeway
- highway
- day/night

- rushhour
- regulations
- winter/summer
- tax
- licence
- repair
- maintenance
- check-up

- repair cost
- garage personnel
- insurance
- driver
- training
- life expectation
- running cost
- maintenace cost

It is now necessary to order and classify these terms and to decide which characteristics, factors, events and conditions must be taken into consideration for the definition of safe and reliable car performance.

As suggested in the preceding section, the failure modes of the car are first investigated and classified, see (Table II):

Table II shows that we are able to define the main terms availability, reliability, safety and production quality, or rather their complements when we know the car's failure modes.

But why do we decide, for instance, that c) "brake fails" is an unsafe or dangerous state? It seems that behind this classification more assumptions concerning the "Environment" are unconsciously made. Namely, the speed may be high in comparison to the available run-out, third-persons and other cars could be endangered, a traffic light may be ahead, etc. Hence, it is necessary to outspokenly define the "Environment" with its sequence of events with respect to time and operation modes or phases in order to clearly define the safety of the "System".

Table II: Failure modes of a car

a) <u>Not reparable:</u>
  - does not start
  - in repair
  - in maintenance
  - etc.
	NOT AVAILABLE

b) <u>Fails during operation:</u>
  - motor fails
  - oil pump fails
  - headlight fails
  - etc.
	NOT RELIABLE

c) <u>Dangerous failure during operation:</u>
  - brake fails
  - loss of a wheel
  - loss of steering
  - etc.
	NOT SAFE

d) <u>Defects, minor failures:</u>
  - early rusting
  - loose interior
  - early rattling
  - etc.
	BAD PRODUCTION QUALITY

Another question is, why (Table II b) the "failure of one headlight" is classified as a system failure, since a car could also be operated at night with the redundant one headlight. The answer is: If the law is also considered, a traffic regulation there says, that a car may not be operated with one headlight. But the redundant headlight would be enough for a safe stop. If one also distinguishes between the failure of the left and right headlight, the failure of the left headlight may be considered an unsafe system failure, if the approaching traffic is taken into account.

These examples show also how particular the terms safety, reliability and availability in contracting and engineering must be dealt with in order to avoid misinterpretation and omissions, and hence design errors and performance failures [5] - [8], [10].

Ordering and classifying the terms of Table I will lead to the more generalized presentation (Fig. 1) of the complex system "car".

SYSTEM EFFECTIVENESS

The connection to the terms cost and overall utility considered over the whole life cycle (table I) is done in Fig. 2. There the term "system effectiveness" is defined by the ratio of utility to cost referred to the total life cycle of a system [8] - [10]. The system effectiveness is thus the most general figure of merit to evaluate system performance, long-term operating behaviour and cost over time, considering all possible conditions. The term "operational effectiveness" is, together with its relationship to the specifications, displayed in Fig. 3.

For the example of the car, a system analysis could result in the systemeffectiveness of an expensive, fast car being low, compared to that of a small, slower car if, according to the definition, the utility of high speed to procurement, operating and maintenance costs is compared, considering a speed limit of 130 km/h or 55 mph (don't use a steam-hammer to crack a nut!).

DEFINING SAFETY AND RELIABILITY BY THE SET OF SYSTEM FAILURE MODES

Generalization of the example in Fig. 1 will lead to Fig. 4. Following the procedure of determining and classifying all the system failure modes in order to evaluate system safety and reliability will lead first to the difficult task of determining all failure modes of each component (hardware, software, etc.). If the relationship between component

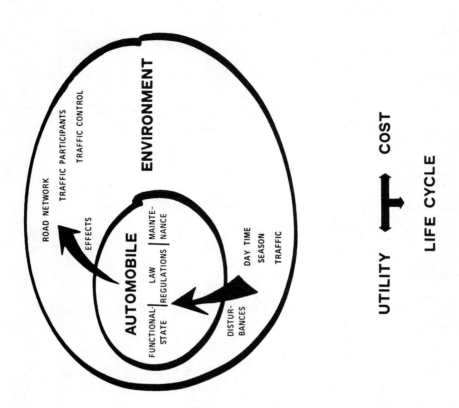

Figure 1

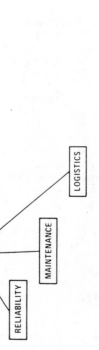

Figure 2

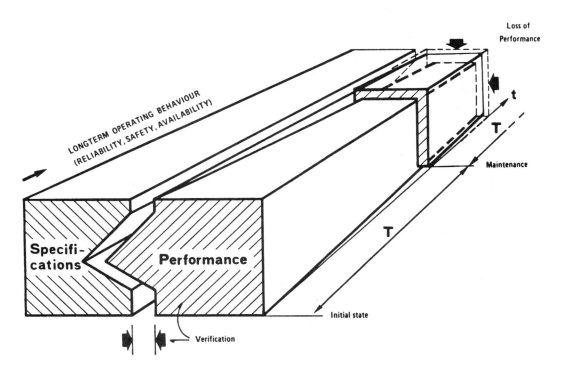

Figure 3

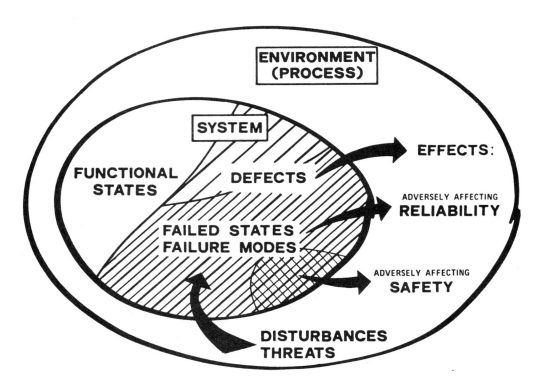

Figure 4

failure modes and system failure modes is established (system structure), each <u>system failure mode</u> is evaluated and rated as to <u>its effect on</u> the safe or reliable operation of the Environment or the Process the System controls (including multiple failures).

While both hardware and software perform a system function, the basic difference between the failure behaviour of hardware and software must be taken into consideration for the system specification, realization and verfication.

Hardware

Hardware failures occur randomly in time, the failure distribution depending on the specific physical properties of each component, its strength and the particular stress applied, as given by its function and the environmental conditions. Hardware failures therefore occur in many modes. Due to some mechanisms a hardware failure may trigger others simultaneously or a certain sequence of failures spaced in time, e.g. [6], [13]-[18]. Hardware components generally age or wear, especially under certain conditions and may thus be cause of common mode failures, e.g. [21], [22]. Once corrected, a hardware failure may occur again. Hardware performance is therefore improved by maintaining the components.

Software

Software errors (not failures [12]) are occurences of fundamental nature. They may be classified as concept, design or logical errors and coding or programming errors. Consequently, software errors are basically not of stochastical nature. But the occurence of software errors, caused by randomly changing process states, operating conditions and modes, as well as hardware failures may cause them to appear to be stochastic. Hence, a software error, once corrected, will not occure any more, Hence software can not be maintained in the sense of hardware, e.g. [18], [23].

As a result, the generalized model presented in Fig. 3 and 4 is in fact a convenient, simple guide for understanding and dealing with complex systems concerning safety and reliability aspects and defining terms in connection with safe and reliable industrial computer systems not defined elsewhere [12].

QUANTITATIVE EVALUATION OF SAFETY, RELIABILITY AND AVAILABILITY

For the quantitative evaluation of safety, reliability and availability mathematical models are used, e.g. [1], [2], [5], [11]

[16], [18]. For hardware the basic data, required are the probability of occurence of component failure modes, component failure rates, frequencies of environmental events, etc. (e.g. [15], [19], [20])

The quantitative evaluation of software is not yet very advanced. However, several models are available, but statistical data concerning software errors is scarce, e.g. [23], [24], [25]

A comparison between reliability and safety characteristics is given in Tabel III, [1] [11].

CONCLUSIONS

Starting point of an engineering project are the system specifications. In order to be a measure, the specifications ought to be a true description of what the system's function is to be and of how the system is expected to perform over a certain length of time considering all environmental and human factors and conditions (Fig. 3).

To assure "correct" performance ("according to specification" [12]), a validation and verification process must be imposed during system planning, designing, production, final testing and the operating phase, i.e. over the whole life cycle. In the context of reliable and safe computer systems engineering and verification of correct performance many questions arise already as to what the terms hardware and software performance, integrity, verification, etc. really mean and comprehend.

In order to support unequivocal, efficient communication and prevention of misunderstanding, misinterpretation and finally failures in system engineering and operation, the Glossary of Terms and Definitions [12] was therefore worked out and the model presented in this paper to materialize reliability, safety and availability of complex systems.

LITERATURE

[1] H. Frey: Computerorientierte Methodik der Systemzuverlässigkeits- und Sicherheitsanalyse. Diss. Nr. 5244, ETH Zürich, 1973.

[2] H. Frey: Safety evaluation of mass transit systems by reliability analysis. IEEE Transactions on Reliability, August 1974.

[3] H. Frey: Die Sicherheit als Bewertungskenngrösse für Automatisierungssysteme. Tagung Technische Zuverlässigkeit, 1975. Nürnberg, No DK 62-52: 65.011.56.

Table III: Figures of merit for system longterm operating behaviour

RELIABILITY — SAFETY

all system-failures modes j * (set F)

Set F of all observable system-failure modes j *

only certain system-failure modes j * (subset $F_u \subset F$)

Repair of passiv redundant components without interruption of operation

$$R(t) = 1 - \sum_{j \in F} P_j(t)$$

$$S(t) = 1 - \sum_{j \in F_u} P_j(t)\, G_j(t)$$

Mean Time To Failure:

$$MTTF = \int_0^\infty R(t)\, dt$$

Mean Time To Safety Breach:

$$MTTSB = \int_0^\infty S(t)\, dt$$

system failure rate:

$$z(t) = -\frac{1}{R(t)} \frac{dR(t)}{dt}$$

rate of safety breaches:

$$z_s(t) = -\frac{1}{S(t)} \frac{dS(t)}{dt}$$

Repair of any component, permitting interruption of operation

Availability of the system:

$$A(\infty) = \frac{MUT}{MUT + MDT}$$

Availability of save operation of the environment:

$$A_s(\infty) = \frac{MTBSB}{MTBSB + MTTRTS}$$

MUT = Mean Up Time

MDT = Mean Down Time

MTBSB = Mean Time Between Safety Breach

MTTRTS = Mean Time To Return To Safety

* see Figure 4

[4] H. Frey: Die Fahrsicherheit als Kenngrösse für die Beurteilung des elektrischen Systems eines Hochgeschwindigkeitsfahrzeuges. 3. Statusseminar für "Spurgebundener Schnellverkehr mit berührungsfreier Fahrtechnik". BRD: Forschungsvorhaben NT 248-250 BMFT, März 1974.

[5] H. Frey, H. Glavitsch, H. Wahl: Availability of Power as Affected by the Characteristics of the System Control Center. Part I: Specification and Evaluation. Part II: Realization and Conclusions. IFAC Symposium, Feb. 21-25, 1977, Melbourne.

[6] H. Frey: Zuverlässigkeitsplanung von Elektroniksystem, E und M Elektronik und Maschinenbau, Heft 6/7, Juni/Juli, 1978.

[7] H. Frey, K. Roth: On the Relationship between Redundancy, Maintenance and Safety Margins for Testing Thyristor Valves. CIGRE Study Committee No 14, August 1978, Paris.

[8] H. Frey: Optimierung der Systemwirksamkeit elektrischer Analgen mittels System- und Zuverlässigkeitsplanung. Vortrag im Rahmen des Seminars "Industrielle Elektronik und Messtechnik" an der ETH Zürich vom 8.11.1978.

[9] H. Blatter, H. Frey: System Effectiveness and Quality Assurance of Electronic Systems. BBC Review, No 11, 1978.

[10] H. Frey, K. Signer: Reliability Planning of Systems. BBC Review, No 3, 1979.

[11] H. Frey: A General Safety Model and the Relation to Reliability and Availability. European Purdue Workshop on Industrial Computer Systems, TC Safety and Security, 1976, Paper No 45.

[12] H. Frey: Glossary of Terms and Definitions Related to the Safety of Industrial Computer Systems. European Purdue Workshop on Industrial Computer Systems, TC Safety and Security, 1977, Paper No 132.

[13] J.R. Taylor: Sequential Effects in Failure Mode Analysis. Conference on Reliability and Fault Tree Analysis, Berkeley 1974 (DAEC, Risö-M-1740).

[14] J.R. Taylor: A semiautomatic method for qualitative failure mode analysis. April, 1974 (DAEC, Risö-M-1707).

[15] Reliability Analysis Center RADC/RBRAC. Reliability Data Books MDR-4÷8, DSR-2, 1976÷1978.

[16] H. Schüller: Methoden zum Erreichen und zum Nachweis der nötigen Hardwarezuverlässigkeit beim Einsatz von Prozessrechnern. Dissertation, 30.10.78, Technische Universität München.

[17] H. Frey: Reliability of non-redundant and redundant digital control computer systems. Proc. of IFAC/IFIP Conference on Digital Computer Applications to Process Control, Zürich, 1974. Springer Verlag: Lecture Notes in Economics and Mathematical Systems; Control Theory Nr. 94 (Part II), p. 500-513.

[18] G. Weber, L. Gmeiner, U. Voges: Methoden der Zuverlässigkeitsanalyse und -sicherung bei Hardware und Software. Kernforschungszentrum Karlsruhe Nr. 01.02. 19p03G, August 1978.

[19] MIL-H DBK 217 B: Reliability Prediction of Electronic Equipment. 1974, updated 1978.

[20] IEEE STD 500: IEEE Guide to the Collection and Presentation of Electrical, Electronic and Sensing Component Reliability Data for Nuclear-Power Generating Stations. 1977.

[21] IEEE Std 323 - 1974: Qualifying Class 1E Equipment for Nuclear Power Generating Stations.

[22] IEEE Std 381 - 1977: Standard Criteria for Type Tests of Class 1E Modules Used in Nuclear Power Generating Stations.

[23] M.L. Shooman: Software Reliability Models and Measurement. INFOTECH State of the Art Conference on Reliable Software, London 28.2. - 2.3.1977.

[24] B. Littlewood: How to measure Software Reliability, and how not to ... . 3rd Internat. Conf. on Software Eng., May 10-12, 1978, p.37-45. IEEE Cat. Nr 78 CH 1317-7c.

[25] S. Bologna, W. Ehrenberger: Applicability of Statistical Software Reliability Models for Reactor Safety Software Verification. Comitato Nationale Energia Nucleare, RT/ING(79)1.

EXPERIENCE WITH COMPUTERS ON SOME UK POWER PLANTS

B. K. Daniels, A. Aitken and I. C. Smith*

National Centre of Systems Reliability (NCSR), United Kingdom Atomic Energy Authority, Warrington, England
*Reactor Development Laboratory, United Kingdom Atomic Energy Authority, Windscale, England

Abstract. Computers are being increasingly integrated into the control and safety systems of large and potentially hazardous industrial processes. This development introduces problems which are particular to computer systems and opens the way to new techniques of solving conventional reliability and availability problems. References to the developing fields of software reliability, human factors and software design are given, and these subjects are related, where possible, to the quantified assessment of reliability. Original material is presented in the areas of reliability growth and computer hardware failure data.

The Report draws on the experience of the National Centre of Systems Reliability in assessing the capability and reliability of computer systems both within the nuclear industry, and from the work carried out in other industries by the Systems Reliability Service.

Also in Appendix 2 some views from the operating experience are represented in the shape of problems that were noted and solutions found.

Keywords. Computer control, reliability data, power station control, operating experience.

INTRODUCTION

Over the last three decades computing and computer systems have evolved from a scientific laboratory environment to many and diverse commercial, military and industrial application areas. As the techniques used to engineer the computer systems have developed, increasing computer power has become available for the same capital outlay, whilst other computer facilities have become available at lower and lower cost. The increased computing power has been used in the solution of more difficult problems, and to provide alternative means of obtaining established problem solutions in a shorter time period. Lower cost computing has allowed the spread of computer systems into areas of business and industry where cost threshold effects are significant.

In considering the reliability of a computer system, the application area of the system must play a large part in establishing both the need and the degree of reliability required of the system. Also, the measures taken to protect the user and surrounding environment from the effects of computer system failure, when it eventually occurs, have some inter-relationship with the design features of the system. The second section deals with this problem under the general heading 'How right must the system be?'

The computer system design is then discussed in the next four sections. 'Static design techniques' are defined as those techniques which aim to improve reliability by the ordered structuring of the system. This section places emphasis on the hardware components of the system and gives examples of quantified analysis techniques which can be applied to the hardware. 'Dynamic design techniques' highlights the features of systems which lead to automatic detection of errors in the running system, and to taking timely and appropriate action. These design techniques have application to both software and hardware. 'Redundant software and other software considerations' deals briefly with the particular problem area of design and implementation of computer programs and 'Modularity', discusses the merits of modular design software, and presents quantified justification of this technique.

The view of a computer system taken by the user is reviewed in the section 'Man-machine interface'. The section refers to systems which have significantly failed to meet the needs of the users. Further examples are given of systems which aimed - and to a large degree succeeded - in providing the appropriate level of interface for the user. These special features are highlighted.

Later sections return to the quantified assessment of computer system reliability.

'Reliability growth' presents data confirming the often observed characteristic of complex systems, that in the early life of the system there is a reduction in failure rate with time. Examples of software, and the associated hardware demonstrate the close relationship between parts of a computer system and the need to consider these interactions. A typical model of reliability growth in a hardware component is given in 'data sheet' format. 'Steady state reliability' is considered in the final section, and typical data for minicomputers and their peripherals are presented. These data are of use in assessing the performance of like computer installations and represent in excess of 100 years field experience.

The report concludes with a brief summary, a list of 22 References, and in Appendices, a definition of the terms used in collecting mini-computer data and a user view of experience with a computer in a nuclear reactor control system.

HOW RIGHT MUST THE SYSTEM BE?

In the analysis of systems many questions must be asked, but when considering reliability, then an important question to be asked is "How right must the system be?" In other words, are we dealing with a home built microcomputer kit, a one off, having virtually no influence on the world around it?; or, do we have a system where there is potential for loss of life, complete destruction of large industrial plant and maybe severe damage to the environment. Safety is one aspect which must be included in the deliberations. Another dimension is the relative cost of the system versus the benefits in normal operation, and also the potential cost of any failure within the system. We may have to consider politics, legislation, personnel relations and public awareness of the benefits and hazards of the project. The designers, maintainers, users and owners of a plant computer control system must find their own answers to these questions.

For some projects it may not be worthwhile to consider other than the simplest or lowest cost system. For other projects, it may need to be demonstrated and accepted as satisfactory by all the pressure groups involved in the decision that all possible attention has been paid to the design and operation of the system.

In the nuclear and aerospace industries there has been an awareness of the need to demonstrate the care and attention to detail that goes into these products. If you are a passenger in an aircraft, then you are sufficiently convinced that all the technology surrounding you in the airframe, engines and instrumentation has evolved to give the current good safety record of air transport. Aerospace technology includes observable redundancy which can be justified in quantified assessments. Redundancy can also be 'obvious' to the non-reliability expert consumer, for example, multiple engined aircraft are generally more reliable than those with single engines and this is readily justified by both qualitative and quantitative analysis.

The need for obvious reliability also applies to the nuclear industry, where the instrumentation and control systems are redundant, diverse and hierarchic.

If a computer monitoring and controlling a nuclear reactor runs into problems, and manual intervention fails to restore the situation, a high level and high reliability shutdown system takes over and remedies the situation automatically. Reference to current work in applying computers to reactor safety systems will be made later in the report.

Having decided what type of system we have or need, the many system design and technologies which are available can be reviewed and a suitable implementation selected. Using quantified reliability assessment techniques it is possible to compare alternative strategies and to check for the suitability of a proposed design against agreed reliability targets. This report gives failure rate data for the hardware of mini-computer systems which can be readily applied at the design stage.

STATIC DESIGN TECHNIQUES

Static techniques aim to improve system reliability by the ordered structuring of the system. This includes such techniques as redundancy and diversity of the main operational features of a system. However, additional factors such as software validation, maintenance on-line, diagnosis and calibration features are also included. Equally the use of higher level languages for programming, and the design of the man-system interface are classed as static techniques.

The quantified justification for redundancy is very easy. If there is a single system A which has a probability of failure p over a given time period, then using two such systems in a 1 out of 2 redundant configuration will give an overall probability of system failure over the same time period of p^2. The probability p has by definition a value in the range of 0 to 1. For the example system in Fig. 1, and choosing p as 0.1, then p^2 becomes 0.01, and the reliability of the system is increased from 0.9 to 0.99 by the addition of redundancy. By using 2 out of 3, or 4 out of 5, or other degrees of redundancy, relatively unreliable system components can be configured to achieve the desired system reliability goal – or can they? Increasing the complexity of a redundancy system will improve reliability up to a point, where other considerations such as spurious action or common cause effects outweigh the gain.

There is good evidence that redundant techniques, by themselves, have a practical limit which restricts a system to an overall reliability figure of about 0.999. Factors which can bring about this limiting effect range from natural disasters, eg lightning strikes, (Interupt, Computer Weekly 2 March 1978) to the on-site construction gang, testing to destruction, the cabling of the redundant system, (Nucleonics Week 21 August 1975) or perhaps to the faulty calibration of all the temperature sensors attached to the control system. Common cause factors are well covered in Edwards (1979). The limit does exist, and there are numerous examples of systems which were thought to be sufficiently reliable until redundancy was defeated by a single event which affected large portions of the system and negated the built-in redundancy.

Fortunately there are well developed techniques to improve reliability by use of diversity. In the case of the electromagnetic effects of a lightning strike, non-electrical systems could be used in a redundant configuration with the electrical system. The possibility of simultaneous physical damaging of cabling can be reduced by routing cabling through ducts which are spatially distributed. In a 2 out of 3 system use 3 separate cabling ducts. If the faulty calibration of all temperature sensors can defeat the redundant system, measure pressure, or density or flow, or some other parameter and use this as a diverse input to the control system.

These techniques are well tried in the nuclear and aerospace industry, and are becoming accepted in the chemical and oil industries. It is a matter of time before the techniques become commonplace in the control of systems by computer. The combined use of redundancy and diversity can lead to a reliability target of 0.99999, and this has been achieved in some systems.

Redundancy and diversity bring additional benefits to the operators of a system; with the inclusion of on-line diagnosis facilities it becomes practical to implement fairly sophisticated checking of control system status, and, by inference, the state of the system being controlled. For example, in a heat exchanger there could be a change in characteristics through a build up of deposits on the heat transfer surfaces. When the control system is operating inside the controllable regime, redundant and diverse sensing of temperature, pressure and flow would allow the control system to check against 'standard' relationships between these parameters and report on any deviations. This reporting could lead to early maintenance action, or avoidance of premature action.

Redundancy assists the testing, maintenance and calibration of systems. With on-line facilities available to the maintenance personnel, a 2 out of 3 system could be temporarily reduced to a 2 out of 2 or 1 out of 2 mode, whilst maintenance is carried out on the third system component. However, with this type of facility there is the new reliability factor to be considered - remembering to restore the system as quickly as possible to its design status. It is of course possible to check and report on system status both automatically and manually. The automatic system can feature timed acceptance of its state. For example, an operator message of the form:-

"If you don't restore the temperature sensor in 5 minutes, a controlled shutdown will begin,"

is likely to bring rapid response - provided the message does not occur too often.

Of course, redundancy and diversity have some drawbacks. In one computer network, the NCSR assessed the availability of communications between the different computer systems (Daniels, 1978). Two views of the network were considered. Firstly, the aim of the system design was to give high network availability from each user's end of the system. To achieve this aim the design included redundant transmission equipment, communications computers and diverse telephone links. Some degradation of throughput was allowed, and the main test of acceptability was that a minimum service should be maintained for the user for the maximum period of time. Typical data were used for the communications equipment and an availability figure obtained for each system user. Table 1 lists the failure rates and repair times used in this work. Similar user data have been derived elsewhere, (Worrall, 1976) and this is given for comparison in Table 2. Figure 2 shows a typical block diagram for the communications between location 1 and location 2. The aim of this service was to provide a high speed link between communications computers A or B at location 1 with computer C at location 2. Also a number of low speed lines between location 1 and 2 were required for terminals. These services were implemented using the configuration of equipment as shown, with the design aim of tolerating a single fault in the system whilst still providing each service.

The assessment took into account the mean failure rate of each system equipment, the mean time to diagnose the location of a fault in the system, the mean time to repair the fault and a reconfiguration time to achieve full system status following repair. The working periods of the system were split into daytime, silent hours, weekends and whole years for calculation of availability and a measure of throughput of the system was calculated.

Table 3 gives the results, typical of this analysis. Looking at the system from the user's view, acceptable performance is to be expected from the proposed design.

However, if we now look at the system as a whole, we get a different picture. Figure 3 gives a graph of the estimated probability of a given unavailability being achieved for the complete system. This graph was obtained by simulation of the network using the NCSR computer program PADS (Daniels, 1979). Looking in detail at the graph we can see there is a mean probability that the system will have a fault somewhere for 22% of the time. The symmetrical 90% probability interval indicates that something, somewhere in the full communications network will be faulty between 18 and 27% of system time.

So, for this communications network, the price of high availability for each user of the system is paid in a large maintenance effort. This is revealed in the aggregate view of the system.

A final point to be made on static design techniques is whether to use a single central computer, distributed computers, or hierarchic computers to implement a control system. In the late 1960's single, central computer systems were advocated by salesmen of many systems. Economic forces at that time dictated a single processor and the use of multiplexing techniques for inputs and outputs of the system. However, even at that time, some redundant systems were constructed where the need for high reliability outweighed the cost of implementing the redundancy. Today the economic climate is somewhat different. The relative cost balance between processor hardware, software, control system sensors and actuators can easily favour the use of several computers. Provided the rules inherent in redundant and diverse design are followed, the trend to distributed and hierarchic systems of control computers is encouraging and is likely to lead to improvements in facilities accompanied by the meeting of reliability targets.

DYNAMIC DESIGN TECHNIQUES

Dynamic design techniques are aimed at the automatic detection of errors or faulty situations, and taking timely and appropriate action. The simplest, and frequently used technique, is the parity bit, which will indicate whether there is a single, or odd number of bits changed in a computer storage or transmission unit. Longitudinal redundancy checks, or cyclic redundancy checks, can be added to blocks of computer storage to indicate that even numbers of bits have been changed in the block of storage. Single or multiple errors can be detected by these fairly simple techniques. The action to be taken on detecting an error will depend largely on the timing and location of the error within the system.

If a handshaking interface design is used between the source of the erroneous data and the destination, then a request to repeat the sending of the data will get over any temporary fault or noise interference problem. If repeated sending does not cure the problem then the source equipment may record the error and inform the operations or maintenance team. Receiving conflicting information may lead to selection of a data block by a majority voting system, where for example 3 out of 5 attempts at sending a block of data are indicated as correct.

Where there are time or cost constraints on repeating data, then error correction procedures may be used. Here the parity bit per unit of data is combined with additional redundant bits so that single and perhaps multiple bit errors can be detected as previously, but additional information in the redundant bits can be used to indicate what the error is, and so to correct the error. The code developed by Hamming (1950) is a well known example of an error correcting code, which uses an extra n bits attached to each group of $2^n - 1$ data bits. These extra 'Hamming' bits will give detection of most errors and correction of some. At the bit level of computer data there are many well established techniques for preserving, and knowing the correctness of the data. These techniques are useful in the transport of data without error between parts of the computer system, or between computer systems.

The trend in computer systems is to standardise on methods of communication, and these methods generally include error testing and correction techniques. As these standards are generated, then with the passage of time the communications method becomes implemented in hardware rather than software. As will be mentioned later, this standardisation process can promote an increase in the reliability of systems. When a clear standard is set, then it becomes practical to test for compliance with the standard, and also to accumulate data on the relative performance of different designs. 'Software becomes hardware' is a useful qualitative goal for a general improvement in both the quantification of reliability and the assurance of equipment performance.

There is a further source of data error which can occur both in computer control systems and in commercial data base system. In the control computer environment, are checks ever made that a request for data from <u>Temperature Sensor 21</u> results in real data being received from that sensor? The equivalent problem in database technology is to supply a key for a data record, and to check that the record which is returned is related to the key supplied. The most satisfactory method of solving this type of problem is to add further redundancy to the data. For the temperature sensor it would be practical to return with the data the identity of the sensor, so if we ask for data from <u>Temperature Sensor 21</u>, then in addition to the coded representation of temperature value, we get an abbreviated name of the sensor function (Fig. 4). The sensor function, abbreviated to HTEXIN or in expanded form to "heat exchanger inlet", is returned with the data, all protected by parity and error correction if required.

The computer system can then look up to see if it expected HTEXIN as a response to 'Temperature Sensor 21'. It is important that the response is different to the request, otherwise there is no positive feedback check from the sensor. Gilb and Weinberg (1977a) propose the name "Checkword" for this type of facility in database technology.

Two levels of dynamic data security have been described, and these achieve high confidence that the data passed from A to B are correct, and that B is aware that A is really A. A third level of security can be achieved, and this can be implemented as a test that the data received from A are reasonable. Conventionally in multiplexed input to a control or data logging system, some inputs are assigned to test values. By obtaining data from the test channel and comparing with the expected value the correct operation within specification can be validated for any conversion device, eg analogue to digital converter, digital volt meter, servo-digital etc. Similarly, with other types of continuous measurement, or non-multiplexed systems, it is possible to arrange for special coding schemes in the data which will reveal errors in measurement. The Grey Code incremental technique is one such example, and further techniques are given by Johnstone and Pooley (1976) which can be applied to individual sensors.

So, if the system warrants it, and that is a decision to be made for each system, there are techniques which will assure the correctness of data from the control system boundary inwards and similar techniques can be applied to outputs. Errors which can occur due to bad placing of sensors, or poor action of control valves for example must be controlled by applying the accumulated engineering experience in those areas. Feedback of information from additional sensors on control actuators is frequently used to provide indication of faulty operation.

Potential errors in data appear to be controllable, but does this apply to the program logic and arithmetic operations? There are several techniques which have practical application in this area, the report covers three briefly. First, there is a developing area of computer software engineering directed to fail-safe design. Keats (1976) covers the design of a nuclear reactor computer based protection system. The correct operation of the system is indicated by a cyclic sequence of binary 1 and 0 states. If the system fails, then the cyclic sequence is interrupted, resulting in a continuous stuck at 1 or stuck at 0 state. If the computer hardware or software fail then this leads to corruption of the cyclic sequence. Data are cycled through the computer memory, using different memory locations, and ensuring all bits of each memory location are cycled through 1 and 0. Keats concludes that it is possible to design a fail-safe system provided that each bit of every memory location of the system is regularly exercised at a rate determined by the maximum repair time of the system, and the repair time must be short compared to the mean time between failures. That is, a system with inherent high availability may be made fail-safe using cyclic techniques.

The second method of checking program logic is being implemented in the hardware of the computer systems. In idle times, for example waiting for a clock signal or external event, the computer is designed to cycle through a program which tests the correct operation of each instruction (Savarez, 1978). This is an extension of the load-time diagnosis facility, where immediately after switching on the system power, diagnostic programmes are executed before the computer is available to user software.

A third method of checking the paths taken by user software has been used. Each software component changes a "routing" variable, and on completing a cycle of operations, this routing variable is checked for compatability with the values which would result from the designed software paths. You and Cheung (1975) have covered these techniques under the generic title "Self-checking software". They indicate the possibility of locating software errors dynamically, and of using similar techniques to assist in the verification of the software and recovery from error situations.

These software techniques are in the early stages of development, but are being used increasingly in systems where high reliability is essential.

REDUNDANT SOFTWARE AND OTHER SOFTWARE CONSIDERATIONS

A current controversial topic is the design and use of redundant software. The proposition is that if sets of software performing the same function are implemented using different algorithms, then the results of the software can be compared in m out of n type voting arrangements and true software redundancy has been achieved. The majority voting technique proposed for improving hardware reliability can then also be applied to reap equivalent benefits in the software Gilb (1977b) and Avzienis (1977).

If as Seaford (1975) proposes, separate programmers are used for each redundant software module, then a degree of independence is built into the overall system design. Errors in one implementation are unlikely to be exactly matched in the performance of other modules. If we consider modules as data handling packages then the correctness of the module can be judged by the verification of the data. If the testing of redundant software reveals significant differences, then they can be investigated at this stage. Also, if redundant modules are coded in different computer languages, then a degree of protection against some of the characteristic errors of these languages can be achieved. A characteristic FORTRAN error is exemplified by

$$ERORR = ERROR + 1$$

where ERORR and ERROR will be two different variables. Without additional information it is not possible to say which variable the programmer intended to increment. By coding in other languages, perhaps ALGOL-like, this mistake would be highlighted since ALGOL-like languages require the declaration of all variables. In using redundant modules we may wish to exchange efficiency of code execution for increased reliability in the software. Gilb (1977) quotes an example of coding first in the language APL for efficiency in debugging and avoiding the need to flow diagram the software. Then recoding into FORTRAN for efficiency in run time processing.

There are many high level language preprocessors which aim to make easier the life of the applications programmer. Commonly used features are given high level named calls, which are automatically converted into thoroughly checked COBOL, FORTRAN, real time language routines etc. The arguments which support the use of any high level language for applications programming apply with at least equal relevance to higher level sets superimposed on a standard language. The facilities provided by the computer system become a closer match to the view both of the implementer of a system and its users. The report comments elsewhere on the potential use of database query language structures to aid operators in the problem solving situation.

MODULARITY

One of the claims made for modular computer software is the ease of checking for correct performance. As a design technique, modularity also eases the problems of constructing and maintaining software packages using teams of personnel. There is a suspicion that the problem areas are transformed from writing a large compatible code, to ensuring that not only are the smaller modules correct in themselves, but also the interfaces between modules are specified and implemented correctly.

Work carried out by Endres (1975) on the initial development of the IBM DOS/VM operation system in Germany gives quantified support to some of the aims of modularity. The DOS/VM operating system at that time was composed of 422 modules varying in size from a few hundred statements to several thousand. During the design and implementation of the system some 432 errors were encountered, recorded and classified. At the design stage, extensive checking of the program specifications was carried out by separate teams. Some checks used formal description and proof techniques to compare the design with specification. Other work used simulation testing to validate design data and techniques. Altogether this work detected about half of the total errors encountered.

The remaining errors came to light in test runs carried out by the system programmers and by independent inspection of the detail coding of the modules. This figure is of primary interest when considering validation techniques and costing.

Table 4 apportions the cause of system errors to groupings of modules. It is interesting to note that 85% of the errors required changes to only one module per error, and that only 1.5% of the errors required changes to 4 or more modules. In correcting the most complex error, 8 modules were changed.

Viewing the system from the modules aspect, Table 5 shows that just over half of the modules were error free. Of the modules contributing to errors, about one quarter had a single error and these made up 22% of the total errors. Nearly two-thirds of the error were caused by only 20% of the modules, and a further 12% of the errors came from only 3 modules. These most error-prone modules were also the largest having more than 3000 statements each. The Endres data are perhaps sufficient justification of the use of modular design to control and isolate errors. If a module (or indeed program) grows too large, and a suitable guide figure may be 2 to 3000 source statements maximum, then the number of errors that occur is likely to increase rapidly. There is nothing in the DOS/VM data which would indicate that the converse applies, that is limiting the size of modules to a very small number of statements will not offer any guarantee of low error rate. There is perhaps a threshold effect.

As can be seen from Table 5, a single module contributed 3% of the errors. This was considered to be an "unstable module", and the degree of error was not unexpected. Perhaps the real difficulty of using modular coding schemes is not in specifying the interface between modules, but in ending up with one module which has to handle the unsolved problems of all the other modules. A strong human characteristic is to 'pass the buck', but there is an inevitable end-stop to this process. Perhaps, in the implementation of software systems, this leads to a single or small number of 'unstable' modules.

Modularity applies equally to computer hardware. The electronics field is in the forefront of standardisation in the engineering world. It really is quite remarkable how in such a rapidly developing technology engineers can find the time to create standards. One of the major side effects of standardisation is that it becomes much easier to collect data on the failure causes, operational environment and experience, from equipment in the field and combine these with other data. The techniques of assessing hardware reliability characteristics are well established and the NCSR has ongoing programs of data collection for the SRS Data Banks and has also verified assessments against system performance (Snaith, 1978).

Hardware modularity also has applications in the maintenance and construction of systems. Truly 'plug compatible' hardware components make full use of established technology in the interface between components. Maintenance problems can be eased with on-line error detection, providing location and diagnosis facilities down to a suitable module replacement level. The module could be a component, printed circuit board or major sub-system, for example a memory unit, and with power-up replacement the system can be restored in minimal down time with a pre-checked replacement.

Modularity is a useful and practical design feature both at the hardware and software levels of a system. Improved reliability can be demonstrated for both these areas. It is also possible to improve the availability of a system. Reliability improvements will extend the time between failures, which, with fixed outages for maintenance, will give availability gains. By reducing outage time when a fault does occur availability can be further improved. Improved fault detection, isolation and maintenance access assist minimal outage targets. Redundancy can be used to delay the need for maintenance until equipment or personnel are available.

THE MAN-MACHINE INTERFACE

The earlier sections of the report have covered those areas of computer system design which are relevant to the designer, the implementer and to some extend the software and hardware maintenance personnel. The various users of the system have not been considered, except that the overall aim is to provide the user with a reliable system.

What needs to be considered in the computer control system-alert operater interface? Some factors are obvious; when in normal working mode the operator should be convinced that the equipment is working satisfactorily. The details of the interface, control desks, VDU's, log reports should be available in the form and at the time they are required by the operators. But most important of all the operator should be able to tell when he must take over from the machine, and in the system "Human operator mode" the machine should be capable of providing help in driving the plant. For this, the interface needs to be very general. Mistakes have been made in the past by severely limiting the type and format of plant data to just those required in normal operation. Perhaps we should adopt a policy of "worst case information design" when selecting the instrumentation needs of a plant, not just to include control parameters, supervisory or shutdown sensors, but to instrument in a way which will improve the operators 'feel' for the system. We are all happier with independent evidence of an event, rather than just a single report with unknown biases. Design should embrace abnormal system behaviour.

In the field of reliability engineering, the human being is probably the least understood and certainly the least quantified factor affecting systems. Philosophers have qualified man for many centuries. Reliability engineering currently seeks the quantification of man and his effect on systems. The NCSR has been researching into this problem area for several years, and a report on this work is available (Embrey, 1976). There is justified expectation that quantification of human factors is a practical proposition (Embrey, 1977). A strategy has been proposed and further work on collecting data for use in human factor models is continuing.

Voysey (1977) cites instances where the man-machine interface has gone sadly wrong. In the past the revelation that all is not right with the design of the man-machine interface has generally come from the investigation after the event of aircraft and railway accidents. The loss of life in closely controlled environments has led to detailed investigation of the control methods and systems involved. A coroner's court added a rider to a verdict of accidental death on three motorway fatalities that "the motorway signal controlling computer might have contributed to the accident".

An automated steel rolling plant of Royal Dutch Steel has had productivity problems (Voysey, 1977). Two main factors contributed to the low productivity. The operators did not understand the control theory of the computer system, and so tended to let it get on with the job - to the extent that control stations were left unmanned. Also the operators found it very difficult to judge when to take over manual control of the system and when they did intervene, it was too late. They lacked the necessary information, normally obtained by visual contact with the progress of steel through the rolling plant, to be able to assess the effectiveness of the computer system. The designers of the plant had completely enclosed the rolling mill for aesthetics and safety.

These comments introduce perhaps the most important aspect of any computer system the man-machine interface. This interface should aim to be acceptable to all the users of the system at all times. It is not sufficient that the system when working perfectly should perform its required function. The operations personnel must be convinced that the system is performing as required. The maintenance teams must be able to access the hardware whilst it is operating normally to be able to perform standard maintenance, and, of equal importance, there must be access to the system to diagnose what has gone wrong, where, when and why. The software teams should be able to try out fixes, certainly in open-loop form, and preferably also in closed-loop tests.

The key to success in this man-system interface lies in the information which can be accessed by the various users. It is highly unlikely that the hardware people will be at ease with the facilities provided for operators, it could be very dangerous to give operators the facilities which would ideally suit hardware maintenance. The software people will most probably be best served with a third style of interface.

In real time systems one of the most difficult types of problems to investigate is the apparent 'intermittent' fault. This occurs in a non-obvious relationship to other events around the system, and by its nature is a relatively infrequent occurrence. The main difficulty is in catching the problem. Probably the most satisfactory technique is to include in the computer system a diagnostic recording facility. Everything which is carefully selected to be of significance is recorded and remains available for a period of time, say half an hour. So when the 'intermittent' fault occurs, the recorded historical performance of the system can be frozen either manually or, automatically if a suitable trigger sequence feature is included in the system. This type of facility is invaluable to hardware and software maintenance, but has often been added to a system at a late stage, or has only been available off-line.

For the operators, much emphasis has been placed on presenting standard information by VDU, plant diagram, or control panels. Less emphasis is placed on what operators are expected to do in exceptional circumstances. Something starts to go awry, and the standard displays do not present the needed combinations of data. Suitable interfaces have been designed for commercial database applications which allow non-programmers to select the combinations of data they require, and the presentation of these data in report or VDU format. The requirement for operators is to be able to select and view any data on the system to meet the immediate needs. The recommendation here is for the computer control system designer to note the developments in database, query languages and report preparation systems.

The UKAEA developed, in 1974, an interactive control system based on PDP11 computers and the BASIC language (Ellis, 1974). Extensions to the BASIC language were provided such as a MOVE verb which transferred data to control valves and set the speed of gas circulators. Data were accessed via RDA (Read Analogue Data) and RFG (Read Flag) functions. The control algorithms were written in BASIC, each computer controlled two experimental facilities through separate user program areas, and a third supervisory user slot was provided to monitor the control software performance. Two such control systems have been in operation at the Windscale Advanced Gas Cooled Reactor (WAGR) since 1975 and user experience is summarised in Appendix 2.

The aim of this design was to provide very flexible control of the experimental facilities, and to be able to make changes to a control algorithm with minimal interference to other systems on the same computer. Staff responsible for the experiment could readily implement their own control or sequencing algorithms in a high level language, then run and debug on the spot. The control cycle had sufficient delays to allow the use of the relatively slow interpretive language BASIC. But, the user interface to the system was simple, required minimal formal training, and was admirably suited to that environment.

The philosophy of giving access to data was followed up in a later system (1975) installed at Windscale in the new WAGR data acquisition scheme (Ellis, 1976). This system is shown diagrammatically in Fig. 5. Digital equipment PDP11 computers were also chosen for this system. One computer is permanently on-line collecting data from the plant, carrying out alarm scanning and sensor status checking functions. This computer is also connected on-line to an IBM 370 which maintains historical records of the plant behaviour and performs lengthy calculations to determine the correct operational control settings for the plant. The dedicated functions of this on-line computer system used an early version of the RSX11 operating system. The second PDP11 computer system acted as a standby computer for this main supervisory role, and also provided an off-line program development facility. However, during normal plant operation, the second computer was run under a time sharing operating system, which gave users access to the data being collected by the on-line system. Following the success of the BASIC language in control situations, the same language was chosen for the time-sharing system. Plant operators at several independent control stations could develop data analysis programs to suit their individual requirements, and the majority of operational VDU displays were derived from BASIC programs running on the secondary system.

The user interface through BASIC has worked very well in these situations and the design aims have been well met. It is interesting to note that CERN has also chosen an interpretive/interactive language to control the operations of the super-proton-synchrotron at Geneva (Crowley-Milling, 1977). Here the interpretive role has been taken one stage further. Not only is the user-machine interface interpretive as at Windscale, but the machine-machine interface in the distributed computer system is also interpretive. Dr Cowley-Milling also comments that for the

CERN application "The slowness of the interpreter is not a disadvantage – control actions are geared to the operators speed of comprehension, or other equipment to operate".

Where operators are encouraged to take an interest in a computer system there is clear evidence that the system ideology will be adopted and defended against all comers. When operators are further given the opportunity to 'play' with the system and to tune it to their developing requirements, then a much closer relationship with the system is fostered. This often assists in the solving of problems as they occur and operators tend to feel the system is helping them in their work. A content and effective plant operator is a desirable aim, anything which can assist the achievement of this aim should be noted.

RELIABILITY GROWTH

There are a number of statistical models used to predict reliability growth, that is the change in failure rate with time as a system 'settles down'. One model for which there is supporting numerical evidence from both system software and user software environment is the gradual reduction in failure rate with time until a steady state failure rate is achieved. This bears comparison with early and working life phases of the 'bathtub curve' frequently quoted for hardware system components.

Lynch (1975) gives the experience of the first year of user release of an operating system, Chi/OS, at Case Western University, a summary of the data is given in Table 6. From an initial mean failure rate of about 1 per day of operational time, the system achieves a steady reduction in failure rate to 1 per 4 days of operational time averaged over the last four months of 1974. The NCSR has collected data on a user programmed system and these original data are given in Table 7. The data here cover three years operations of the system, and from a first 6 months average failure rate of 1 per 1.5 days of operational time, the yearly average improves to 1 per 5.7 days operational time, and has achieved 1 fault per 6.25 days in the final 6 month period.

There are no similar documented data for the reliability growth of computer systems which have been obtained from field experience. However, a further feature worthy of note in the Lynch paper, is the number of crashes experienced and attributed to the hardware after the introduction of the Chi/OS operating system. A marked reduction in the hardware mean failure rate is shown in Table 6, from 1 fault per 1.9 days of operational time averaged over the first 4 months, to 1 fault per 5.2 days operational time averated over the last 4 months of the reported data.

In the electronic component field the NCSR is developing reliability growth models as a standard feature in the 'Data Sheet' series. Field data accumulated on component failures are being analysed for many electonic component types. A typical data sheet is given in Fig. 6, in this case for medium scale integration TTL logic devices. The data sheet shows how a failure rate may be calculated taking into account the relative complexity of the component, the operating temperature and application environment, production maturity, and the reliability growth factor to be applied to surviving components over a range of operating lifetimes. To quote an example; for a component type 7490, a decade counter, having 43 gates, a mean failure rate of 1.3 per 10^6 hours is obtained for a $40^{\circ}C$ ambient, component application, established production, new off the shelf device. Using sythesis techniques the hardware failure rate of computer systems at any instant in time can be obtained from such component model data. Also, by making use of the component reliability growth models, system hardware reliability growth models may by synthesised.

An assessment carried out by NCSR on a computer control and data logging system which had been in operation for 8 years predicted a mean failure rate for the system hardware of 3 faults per year. This compared well with the measured performance of the system at 1 fault per year over the last 2 years of operation. The computer system was built up from 200,000 individual components, and the assessment allowed for reliability growth.

STEADY STATE RELIABILITY

Data are significantly lacking on the steady state, or long term reliability of computer software, and the NCSR would be interested in any data which can be made available in this area. For computer hardware, the NCSR have conducted surveys of computer systems for several years. Data are particularly rich on Digital Equipment Corporation computers and peripherals, and some generalised failure data are given in Table 8 which covers the years 1967 to 1976. Overall, the Table represents some 115 computer years of reported installation time, with an average utilisation of 70%.

In this Table, the systems are grouped as far as possible by processor design, for example the designations PDP11/05 and PDP11/10 are better indicators of sales market area than of real differences in the hardware. the PDP11/40 computer included in the survey was used as a software and hardware development facility, and generally had a hard life. This undoubtedly accounts for the poor MTTF recorded. All the other systems were used in real time applications, eg control, data logging, communications, etc. The data were obtained from within the UKAEA and also from diverse sources outside. The overall MTTF for all types of computer was 5600 hours. Data on other manufacturers equipment do not vary significantly from this figure.

Table 9 gives in greater detail the basis for Table 8 and indicates the spread of reliability performance which may be expected.

Some of the computer systems surveyed included peripheral equipment and interfaces in sufficient quantity to warrant collecting data. Table 10 gives the data for these equipments. Interesting features in the Table are that VDU's fail 2½ times less often than printing terminals. Paper tape equipment and line printers are on a par with VDU's. Fixed head discs are marginally more reliable than the type of moving head disc on which data were collected. Communications interfaces are comparatively reliable with asynchronous interfaces failing once in 9 years and synchronous interfaces approaching 1 year MTTF.

As is the case with all field derived data, the data may be influenced by the operational environment of the equipment, the maintenance procedures, and the stress placed on the equipment. All the computer and peripherals included in this survey were housed in an office or control room environment. VDU's and printers were used by all grades of personnel. First line maintenance was generally carried out by the resident engineering personnel, and the Digital Equipment Co. contract field maintenance service was used for regular preventative maintenance and on a call-out basis for breakdowns.

The data may be used to predict the performance of similar equipment in similar application environments, but due account should be taken of special circumstances which may affect the performance of any equipment to which these data may be thought to apply.

A list of definitions used in deriving the item descriptions for peripherals is provided in Appendix 1.

CONCLUSIONS

This paper has considered the hardware reliability of computer systems, and original numerical data have been provided which may be applied to the quantified reliability assessment of such systems. The lack of data on software reliability is noted, and encouragement is given to record this important data when it occurs. The computer system is there, why not make use of it to provide this secondary level of data which can be applied to future designs? In the few known cases where information has been collected, it has been used to improve the status of the object system and to provide background data for future work. The poor understanding and research status of the man-machine interface has been discussed and general pointers given to some techniques used to implement successful systems.

The trend to implement well established error detection and correction procedures in hardware rather than software has been noted. The increased use of modular design may ease the collection and establishment of a software/hardware reliability database, and will directly improve the achievable reliability for a given system complexity.

REFERENCES

Avzienis, A and L Cheu (1977). On the implementation on N-version programming for software fault-tolerance during execution. Computer Weekly (2 March 1978), UK.

Crowley-Milling, M (1977). How CERN broke the software barrier. New Scientist, 29 September 1977.

Daniels, B K (1978). The availability of a computer communications network as perceived by a user and viewed centrally. In G Volta (Ed) Nato Advanced Study Institute, Urbino, Italy, Plenum, New York.

Daniels, B K (1979). PADS user guide. National Centre of Systems Reliability, Warrington, UK.

Edwards, G T and I A Watson (1979). A Study of common-mode failures. SRD R146. Safety and Reliability Directorate, United Kindgom Atomic Energy Authority, Warrington, UK.

Ellis, W E (1974). DECUS Symposium, London, UK.

Ellis, W E, J H Leng, I C Smith and M R Smith (1976). The new WAGR data acquisition scheme. TRG Report 2783(W). United Kingdom Atomic Energy Authority, Warrington, UK.

Embrey, D E (1976). Human reliability in complex systems. NCSR R10. National Centre of Systems Reliability, Warrington, UK.

Embrey, D E (1977). Approaches to subjective judgement of human reliability. SRS Quarterly Digest, October 1977. National Centre of Systems Reliability, Warrington, UK.

Endres, A (1975). An analysis of errors and their causes in software systems. IEEE Trans Software Eng, SE-1 No. 2, pp 140-149.

Gilb, T (1977a). Computer Weekly, 15/22 December 1977, UK.

Gilb, T and G M Weinberg (1977b). Humanized Input. Winthrop, Cambridge, Mass.

Hamming, R N (1950). Error detecting and error correcting codes. Bell Syst Tech J (USA), April 1950.

Johnstone, W and D Pooley (1976). The determination and use of sensor status information in computer based monitoring and protection systems. Proceedings of IAEA/NPCC I Specialists meeting on use of computers for protection systems and automatic control, Munich, International Atomic Energy Agency (Vienna).

Keats, A B (1976). A fail-safe computer based reactor protection system. Proceedings of IAEA/NPCC I Specialists meeting on use of computers for protection systems and automatic control, Munich, International Atomic Energy Agency (Vienna).

Lynch, W, J Langner and M Schwarz (1975). Reliability experience with Chi/OS. IEEE Trans Software Eng, SE-1 No. 2, pp 253-257.

Nucleonics Week, (21 August 1975), Browns Ferry Fire.

Savarez, R S (1978). Reliability and maintainability enhancements for the VAZ 11/780. FTCS-8, Toulease, IEEE Catalogue Ch 1286-4C.

Seaford, (1978). Computer Weekly, 2 February 1978, UK.

Snaith, E R (1979). Can reliability predictions be validated. Proceedings 2nd National Reliability Conference, Birmingham. National Centre of Systems Reliability, Warrington, UK.

Voysey, H (1977). Problems of mixing men and machines. New Scientist, 18 August 1977.

Worrall, S (1976). Harware factors: teleprocessing at work. Online 1976, London.

Yau, S S and R C Cheung (1975). Design of self checking software. Proceedings of 1975 International Conference on reliable software. SIGPLAN Notices (USA), 10 No. 5, pp 546-550.

TABLE 1

User experience derived data for communications equipment

System component	MTTF hours	Diagnose/ reconfigure hours	Repair hours	Reconfigure hours
Communications computer	750	0.5	6	0.5
Modem	22000	0.5	8	0.25
Multiplexor	10000	0.5	8	0.5
PO lines	1000 to 4000	0.75	12	0.25
Port switch	100000	0.5	12	0.5

TABLE 2

Comparison data for communications equipment

Multiplexer	MTTF hours
Modem	30000
Multiplexer	16000

TABLE 3

Typical availability and throughput for communications links

		Availability %				Throughput
		Daytime	Silent hours	Weekend	Yearly	
Computer-Computer	Full service	95	93	82	90	95%
	Half service	5	7	18	10	
Low speed	Full service	95	92	81	89	94%
	Half service	5	8	19	11	

TABLE 4

Modules affected by errors, derived from Endres (1975)

Number of modules affected	Proportion of errors
1	85%
2	12%
3	1.5%
4	
5	1.5%
8	

TABLE 5

Contribution to errors by modules, derived from Endres (1975)

No. of errors per module	Proportion of modules	Proportion of errors
0	52%	–
1	27%	22%
2 – 13	20%	63%
14	(1 module)	3%
over 14	(3 modules)	12%

TABLE 6

Operating system crashes from Lynch (1975)

Month	No. of software crashes	Mean failure rate	No. of hardware crashes	Mean failure rate
1	20	1 per day	14	1 per 1.9 days
2	14		15	
3	13		11	
4	12		6	
5	9		9	
6	9		7	
7	9		6	
8	8		11	
9	7		5	
10	6		7	1 per 5.2 days
11	6	1 per 4 days	2	
12	2		2	
13	6		6	

TABLE 7

User system crashes, NCSR data

6 month period	Mean failure rate
1	1 per 1.5 days
1 + 2	1 per 1.7 days
3 + 4	1 per 2.7 days
5 + 6	1 per 5.7 days
6	1 per 6.25 days

TABLE 8

Grouped computer system faulure rates

Processor type	Number	Cumulative years installed	Utilisation = Switched on time / Installed time	MTTF hours
PDP8I	1	9	100%	3410
PDP8L	3	21	51%	4280
PDP8E	2	11	27%	5480
PDP11/05, 11/10	12	23	100%	7420
PDP11/15, 11/20, 11/21	12	46	7%	4080
PDP11/40	1	2¼	25%	1250
PDP11/45	1	2¾	57%	6304

TABLE 9

Detailed breakdown of computer system failure rate

System	Processor type	Memory size	Delivery date	Uptime start date	Uptime hours	Proc faults	MTTF hours
1	PDP8I	4K	1967	1/11/72	30672	9	3408
2	PDP8L	4K	Mar 1969	1/11/72	30672	2	
3		4K	Mar 1969	1/11/72	8216	8	4282
4		8K	1970	1/11/72	8216	1	
5	PDP8E	4K	1971	1/11/72	8216	2	5477
6		8K	1970	1/11/72	8216		
7	PDP11/05	8K	May 1973	2/ 5/73	25510	6	
8		8K	May 1973	2/ 5/73	25510	2	
9		8K	May 1973	2/ 5/73	25510	12	
10		8K	Jun 1973	1/ 7/73	24090	2	
11		8K	Nov 1973	1/12/73	20710	2	7100 hrs
12		8K	Oct 1974	1/11/74	12050	1	
13		8K	Nov 1974	1/12/74	11950	0	
14		8K	Nov 1974	1/12/74	11950	0	
15		8K	Dec 1974	1/ 1/75	11450	2	
16		8K	Dec 1974	1/ 1/75	11450	0	
17	PDP11/10	8K	Nov 1974	1/12/74	11950	2	11700 hrs
18		8K	Dec 1974	1/ 1/75	11450	0	
19	PDP11/15	8K	Jan 1972	1/ 1/73	28470	1	
20		8K	Mar 1972	1/ 1/73	28470	4	
21		8K	Mar 1972	1/ 1/72	18470	1	13733 hrs
22		8K	Mar 1972	1/ 1/73	28470	0	
23		8K	Jan 1973	1/ 2/73	23710	3	
24		8K	Jan 1973	1/ 2/73	27740	3	
25	PDP11/20	16K	Oct 1971	1/11/71	20152	12	
26		16K	Oct 1972	1/11/71	17490	15	2041 hrs
27		8K	Mar 1972	1/ 1/73	23710	6	
28		8K	Mar 1975	1/ 4/75	8030	1	
29	PDP11/21	28K	1972	1/11/72	23600	7	2052 hrs
30		28K	1972	1/11/72	23600	16	
31	PDP11/40	16K	1974	1/ 2/74	5000	4	1250 hrs
32	PDP11/45	24K	1973	1/ 6/73	13607	2	6304 hrs

TABLE 10

Computer peripheral equipment data

Item Description	DEC type	Quantity	MTTF hours	Comment
Teletype	ASR33 ASR35 ASR39	12	1060	
Dot matrix terminal	LA30	1	450	30 charac per second keyboard/printer
VDU	VT05	8	2640	
Paper tape punch/read	PC11	5	2790	300 charac per second reader 50 charac per second punch
Card reader	CR8	1	8200	300 card per minute
Line printer	LS11 LS8	5	3100	Dot matrix 165 charac per second
Asynchronous interface	KL11 DL11 DC11 KL8M	30	81000	
Synchronous interface	DP11	4	7800	
Fixed head disc	RF11	2	2800	512K byte disc
Moving head disc	RK05	5	2200	2.4M byte each drive

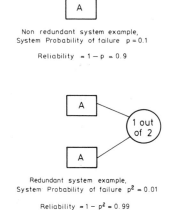

Fig. 1. Example of Redundancy with Quantified Reliability.

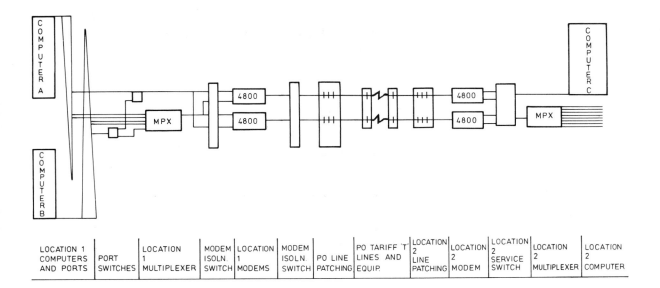

Fig. 2. Example of Computer Communications Equipment Between Locations 1 and 2.

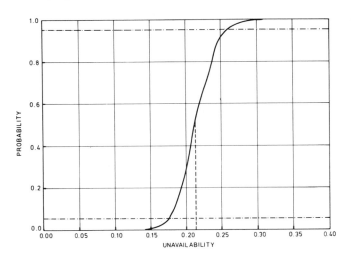

Fig. 3. Complete Computer Communications System, Central View.

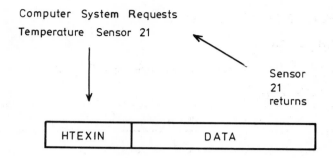

Fig. 4. Example of Data Checkword

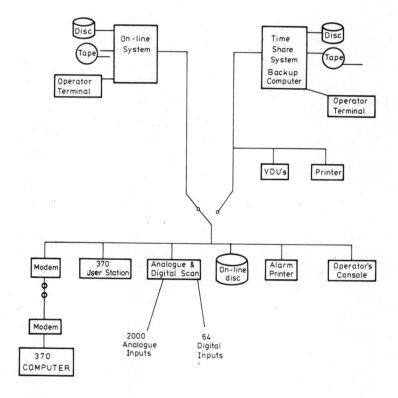

Fig. 5. Block Diagram of the Wagr Data Acquisition System.

DESCRIPTION   Microcircuit, digital, TTL, SS1/MS1

ITEM CODE   54670.57.86.0.00.0.0

Operational failure rate $F = (F_e + F_t)K_1 \cdot K_g$ failures per 10^6 hours

Storage failure rate $F = 0.7F$

Values of F_e									
Application code	Number of gates in device								
	1	2	4	6	10	16	30	50	99
1	0.0047	0.0060	0.0076	0.0088	0.011	0.013	0.016	0.019	0.024
2	0.0070	0.0090	0.011	0.013	0.016	0.020	0.023	0.029	0.036
3	0.012	0.015	0.019	0.022	0.027	0.033	0.039	0.048	0.06
4	0.023	0.030	0.038	0.044	0.053	0.066	0.078	0.096	0.12
5	0.047	0.060	0.076	0.088	0.11	0.13	0.16	0.19	0.24
6	0.094	0.12	0.15	0.18	0.21	0.26	0.31	0.38	0.48
7	0.12	0.15	0.19	0.22	0.27	0.33	0.39	0.48	0.60
8	0.19	0.24	0.30	0.35	0.43	0.53	0.62	0.77	0.96
9	0.35	0.45	0.57	0.66	0.80	0.99	1.2	1.4	1.8

Values of F_t									
Temperature ambient °C	Number of gates in device								
	1	2	4	6	10	16	30	50	99
0	0.0004	0.0006	0.0010	0.0013	0.0019	0.0031	0.0057	0.011	0.017
25	0.0013	0.0021	0.0035	0.0044	0.0063	0.010	0.020	0.037	0.058
50	0.0042	0.0068	0.011	0.014	0.020	0.033	0.062	0.11	0.17
75	0.012	0.019	0.031	0.039	0.056	0.092	0.17	0.28	0.43
100	0.028	0.045	0.073	0.093	0.13	0.22	0.41	0.63	0.97
125	0.060	0.097	0.16	0.20	0.29	0.47	0.88	1.2	1.9
150	0.12	0.19	0.31	0.39	0.56	0.92	1.7	2.4	3.7

Values of K_g (Reliability growth factor)	
Age hours	K_g
100	15.0
1,000	3.0
10,000	1.0
100,000	0.4
Ultimate	0.16

Application codings	
Code	Application
1	Digital computers.
2	Data handling systems.
3	Control rooms, laboratories.
4	Normal industrial, ground based.
5	Shipboard, chemical environments.
6	Mobile, road, rail.
7	Portable and bench applications.
8	Airborne.
9	Rocket Launch.

Failure mode	Factor
Output high (1)	0.5
Output low (0)	0.5

$K_1 = 10$ for new production.
$K_1 = 1$ for stabilized production.

If required, assume 4 transistors per gate.

FIGURE 6   TYPICAL 'DATA SHEET'

APPENDIX 1 LIST OF DEFINITIONS

CPU   Arithmetic Units
      Code Processor
      Memory   Core   RAM
                     MOS   ROM
                     Bipolar Firmware
      Operator's console

<u>Asynchronous Interface</u> Connects a CPU to a (usually) low speed device using asynchronous serial transmission techniques. Usually in the range 50 to 9600 bits per second (bps). May be connected to a modem to give access to the CPU from a remote device.

<u>Synchronous Interface</u> Connects a CPU to a (usually) high speed device using synchronous serial transmission techniques. Usually from 600 bps upwards. May be connected to a modem to give access to the CPU from a high speed remote device.

<u>Teletype</u> Low speed (usually 110 bps) asynchronous device that will commonly provide a keyboard/paper tape reader and a printer/paper tape punch configuration. For example the Teletype Corporation models ASR33, ASR35 and their equivalents. Requires an asynchronous interface to connect into a CPU.

<u>Dot Matrix Printer</u> Low speed printer perhaps 10 characters per second up to 200 characters per second printing speed. May be combined with a keyboard to give a terminal which can replace the teletype. The printing action is by needles impacting the inked ribbon and forming an impression on the paper, each character will be formed from a matrix of dots commonly 5 x 7, 7 x 9 or upwards.

<u>Line Printer</u> A printer which commonly inputs as data a string of characters and then prints that string on a single line. Teletypes print a character at a time. Normally the Line Printer will include facilities for handling various widths of paper, lengths of pages and also programmable vertical format control. May use dot matrix or other techniques for forming characters; generally runs at 120 to 12000 lines per minute and may run at different speeds depending on line length and character set in use. Includes the interface with the CPU.

<u>Paper Tape Reader/Punch</u> Reads 5, 7 or 8 hole paper tape at 50 to 1000 characters per second, may accommodate different tape dimensions. Also as a separate device will punch paper tape at 20 to 100 characters per second. Includes the interface with the CPU.

<u>VDU</u> Vidio display unit; commonly combined with a keyboard and used instead of a teletype in those applications not requireing hard copy. May be direct replacement for a teletype, run at higher speed, or even incorporate features to edit information on the screen before transmission. May include the facility to output the screen data to a local printer. Requires either an asynchronous or synchronous interface to connect with CPU.

<u>Magnetic Tape</u> Drive units and interface to CPU to allow reading and writing of data on magnetic tape media. Can take many forms from cassette tapes which may replace paper tape equipment, pseudo-random access tape systems to sequential multi-channel inter-computer media and file compatible systems.

<u>Fixed Head Disc</u> Usually a magnetic recording medium revolving at high speed with a read/write head per recorded track on the disc. Has a latency of 10-20 ms and negligible track access time. Sometimes used for lower than memory speed random access data. More often in time sharing, real time, multiprogramming, or – virtual machine environments will be used for paging or overlaying main memory with currently required data or program segments. Usually in the size range 32K bytes to 10 M bytes per drive.

<u>Moving Head Disc</u> As for fixed head discs except that only one head per recording surface is used which reduces the cost of a system but increases access time to a track. Typical head movements 10 mS per track 70-200 mS average track (seek) movement. Size range per drive 512K bytes – 400 M byte. Covers wide range of devices from low cost/low speed/low volume to high cost/fast/bulk storage.

APPENDIX 2 A USER VIEW OF EXPERIENCE WITH A COMPUTER IN A NUCLEAR REACTOR CONTROL SYSTEM

Introduction Two of the systems discussed in the main paper was looked at in detail to obtain the user view of experience with the computer system. It should be stressed that the computer system considered is not within the reactor safety system proper, but has some related functions.

The aim in recording this experience is to give guidance to designers and to allow assessment of safety, reliability and availability aspects in order to judge whether computer systems on reactors can meet the requirements which might be placed on them, eg, by a licensing authority.

Two of the systems on the reactor are considered in detail. First, the auto control system which is used on experimental facilities in the reactor and the second a system (NUDAS) which collects data from the normal running of reactor as a power station and provides essential alarm scanning.

It was preferable to begin with the hardware side of the auto control system and in discussion a view was expressed that, with a control algorithm that works, there would be no subsequent software problem. In some ways hardware faults have been mitigated by additional features in the software and it was thought that any software failures could be considered as design failures. Protection against some hardware faults had subsequently been incorporated in the software.

Circulator Speed Change The original design called for a step change of any magnitude up to the maximum value of a counter. But in reality a large step change output from the control program demanded an acceleration too great for the circulator to respond. The change made was to modify the controller hardware so that it must complete a demanded change in speed in a number of basic smaller steps rather than one large step. This change was to guard against the possibility of a wrong value being sent out. Software protection was also built in to limit the step size which could be demanded. It was recognised that if a register failed in the hardware sense it might give the wrong setting for circulator speed. It was noted that there was already software protection in the assembler language to limit the maximum size of step change but although the design met the specification, the specification itself was wrong. So far as the circulator itself was concerned, it could have met a higher rate but the decision to get over the difficulty meant that a slower rate for the circulator was operationally the result of changes made. The main concern in the situation was if the register failed when reading back the circulator control state. It was possible before the modification to give out information which was based on wrong data. It was never necessary to invoke the maximum acceleration possible and it was thought prudent to remove its accidental occurrence, because selected accidentally it could result in undesirable side effects.

Power Supplies Despite the fact that power supplies were duplicated, there had been a series of power supply troubles which emanated from the low voltage and stabilised power supplies to the circulator interface. There were internal faults in the power supplies as well as in the isolation diodes and such faults seemed to occur in batches. At one stage it was thought that the isolation diodes were causing the problems, but with all the modifications introduced, components may have been overheated at some time or operating outside design limits in some other way. These faults were later linked with faults in the valve position control and it is now felt that as in the last two years there had not been one fault in this control, either the design faults have been fully corrected or components that were on the edge of failure had been replaced. It was recommended that for future operation the power supplies should be mounted so that they could be got at externally. Although low voltage dual power supplies were provided, it was impossible to work on the faulty one with mains power supplies live. Thus maintenance access for the valve controller was poor.

Valve Controller It is noteworthy that several reactor trips had occurred before any intermittent short circuit fault had been found. The fault was caused by two boards being too close together and a sharp protruding pin on one board punctured the insulation to a capacitor on the other. This obscure fault led to quite a number of false trails, with early blame being put on the software design until the real fault was actually found. The opinion was expressed that this sort of failure could perhaps have been avoided at the design stage.

Failure of Output Drive Transistors The output drive transistor to the slave relay controlling the circulator drive current failed on short circuit, holding in the relay and providing a continuous demand for change of speed. Eventually a reactor trip had occurred. In 5 years operation of eight such circuits, only 1 transistor failure occurred.

Software Faults A fault had occurred which was traced to a memory fault senstiive to two instructions next to each other and this happened within the executive program. The two instructions were operating on a particular area of memory consecutively before the fault could occur. It took three people about 4 days to find the fault which was in effect producing an illegal instruction. This type of fault is not normally found by normal maintenance software testing and in this particular instance, because of the area in which it was located and the particular combination of instructions required, the net result would have been moderately harmless but this does not mean that all such faults would be harmless. It was considered that the classic fault which might be harmless would be in a program continuing to return a constant value whilst in reality many things were taking place on the system. Under these conditions you would not know about any fault that existed. There are then two sorts of problems with errors and instructions. The first modifies the logical flow of the program and the second modifies the data flow. The question of detecting logical flow errors in the program itself could result in data being obtained which does not represent what is really happening. One difficulty was that a lot of the software was written in BASIC It was therefore dependent on the BASIC functions behaving correctly. If the program code is modified just sufficiently to make the behaviour of these functions incorrect, there is no way you can detect that at the BASIC program level. The result is that thorough routine testing of the operation of the software is the only way of getting round this kind of problem.

This kind of situation is typical of the difficulties that arise in attempting a logical analysis of how reliable software is going to be. Problems in producing reliable software and being able to quantify how long the system will work makes estimating the reliability of software a very difficult but not impossible task. It however is relatively easy to demonstrate system reliability for 99% of fault conditions. The operators expressed the opinion that you can have reliable software provided you have reliable hardware and in the system considered all the routes through the software and error routes had probably by now been fully tested.

Therefore the chances of a fault reported on the machine, as it was now operating, being due to a software fault, are minimal. In the complex interaction of hardware and software, trouble is always in the memory, unless some system such as a double memory with parallel type processing or a 2 out of 3 system is used.

One of the operating experiences was that each time a specific example of a fault was brought up, a check for that particular example could be found. It is more difficult to do an all embracing check and even if you had such a system the difficulty would be to have space to introduce it into the system. There were limitations on the auto control program and also on the devices you could use, because it would take too long to check out fully the complete control cycle.

Considering the question of duplication of interacting data, it was felt more necessary to duplicate some key parameters. It appeared that instead of complete redundancy some diversity in the data presented might be preferable. In the system under discussion, 5 or 10 analog inputs could be selected which would indicate that something vital was going wrong and these inputs might represent only some 5 or 10% of the total number of signals being scanned for the auto control function.

In safety terms there is a hierarchy in the auto control system because it is over-ruled by the analog trip system through the safety circuits. It was still felt that such a system would be necessary and should be completely independent from auto control functions.

Data Logging System There were two ways in which a faulty signal could be revealed. First, via the alarm printer which rings a bell to alert the operators. The second is from the transmissions down the link to the 370 computer and it was felt that for both, a number of test inputs with constant voltages would be desirable. A block of 100 with a test point at the beginning or end of each block was considered to be sufficient.

The question of synchronisation between the data logging, scanning and the auto control cycle system was discussed and the query was raised as to whether this too should be heavily interdependent or asynchronous. No experimentation had been done on desynchronisation. The system runs with a pre-set time interval allowing data to come partly from one scan and partly from another, assuming that scanning speed is fast enough. It was not clear whether the system would run just as well synchronised or desynchronised. By running the auto controller and the monitor slot asynchronously you can apparently get the auto control running in the opposite direction to the sign of the error and this led to confusion. Synchronisation is used to freeze a picture of what has happened in say 5 seconds, and it is frozen for a length of time sufficient to read out the data in a meaningful way. It was clear that the operators relied on the monitor slot.

Initially there had been a lot of bugs in the BASIC language system. There were also many problems with the I/O on the system which needed to be corrected. It was calculated that at least 30 or 40 percent of the time was taken in bringing the package up to a satisfactory state. Once this was achieved, very few changes were necessary. The BASIC software has been operating stably since October 1974 but it was acquired initially in 1972.

Fault Reporting It was felt that it would be desirable to have some means of recording what is going on in the processor and questions arose as to whether it was necessary to go into the depths of hardware and record every signal or if recording should take place at a higher level. These records could be used as an aid in fault diagnosis. It was felt by the operators that the automatic recording of data could be overdone and the opinion was expressed that a programmable system was better. There was a decided trade-off between core space, money etc. There are disc errors which it is important for the Systems Manager to notice but it would not necessarily be desirable to have everybody in the control room alerted when something like that happens. The operators in the control room should have their attention on things that are happening on the alarm scanning side but, of course, it was desirable to know very quickly if the whole system had crashed. To this end a routine was set up that would print out a lot of extra information if the system crashed. The important lesson was that information should be directed to particular sets of people at a proper level. Basically, error messages were divided into 3 categories.

1. A print out.
2. Ringing a bell.
3. Messages which represent something going wrong.

There is a further category of messages which can be used as a diagnostic tool and a back history data facility provided some advantages for this. One of the early problems was that there were so many alarms coming in that the system ran out of space to store the calls which had to be stripped off. A method was devised of backing off the messages on to disc store if there were more than a certain number. This would allow the messages to be looked at in the course of time and give the system a chance to catch up but at maximum alarm message rate the lag could get so bad that something like 10 minutes could elapse between generating an alarm message and someone actually seeing it. (It should be noted that operator action should have occurred on the arrival of the first alarm message).

The use of read only facilities and write facilities protected by key locks was probably the best way of running a system with safety and security and it was felt necessary to have isolation of user functions.

The question of intermittent faults proved to be something that remained a problem. Obscure faults, even for widely used software, can occur even after many years of experience, and it appeared that there is probably no way that single purpose software is going to be absolutely bug-free. There can be some dangers in altering a software package. The conclusion was that if something goes wrong in the computer system as it is operating now, it is assumed that it is due to a hardware fault whereas in the past it may have been thought of first as a software error.

Documentation With regard to documentation, it was felt that this had reasonably been kept up to date with modifications but initially there had been some lag between the carrying out of modifications and amendments to the documentation. It was noted that on some systems where data processing is part of the safety system, there is insistence which forces the documentation to be brought up to date before the modification is done but it was recognised that even this is not a perfect system. A number of problems resulted from software being designed at the same time as the hardware it was trying to drive. Even with standard hardware, it may be necessary to combine equipment from different manufacturers and if constraints are placed on software development and modification, the only degree of freedom open is to modify existing hardware or use purpose built hardware to solve that particular problem.

DISCUSSION

Ehrenberger: I have a question on Mr. Daniels' paper. Does the data obtained from operating experience show that software failures occur more frequently than hardware failures.

Aitken: Initially the experience of typical computer systems is that software failures are more frequent than hardware failures. This situation tends to reverse in later life when hardware faults are more frequent than software. At the same time both hardware and software exhibit reliability growth such that failures become less frequent as the system matures. Where correct maintenance is carried out the hardware failure rate tends to stabilise. But, occasionally, an increase in hardware failure rate can be attributed to the wear-out of mechanical equipment in the computer system which is not replaced in time.

Ehrenberger: What portion of the data could not be attributed to either software or hardware failure.

Aitken: Where possible the cause of a failure was investigated and attributed to hardware or software. However, there are inevitably a number of failures which cannot be clearly classified as hardware or software. For example the failure quoted in Appendix 2 of the paper where the occurrence of two adjacent instructions in a particular area of memory resulted in a failure which caused poisoning of the software; which is cause, which effect, in this example. We cannot quote proportions of unattributed failure causes.

Lauber: Could you give an estimate?

Daniels: There is considerable difficulty in obtaining data on the failure of computer systems and the subsequent analysis on the cause of failure. This applies to both hardware and software. For example, Appendix 2 cites an example where initially the software was suspect but eventually the fault was traced to hardware. It requires very good record keeping by both software and hardware teams of faults and repair action taken. The NCSR has acquired data where this has been available.

Sometimes this has been for hardware only, less frequently for software only and never for both software and hardware in sufficient depth to answer the questions that are now being asked. This further emphasis the need for recording of failure data by designers, operators and maintainers of computer systems and to allow access to this data.

Gilbert: In the systems described in the paper both computers were located in the same control room. Doesn't this lead to problems, for example when there is a fire?

Smith: In the particular case quoted in the paper, a separate control room was available for emergency situations. Complete back-up of all systems was provided (not using computers) which would handle this and similar events.

GUIDELINES FOR THE DOCUMENTATION OF SAFETY RELATED COMPUTER SYSTEMS

A. A. Levene

Systems Designers Limited, Systems House, 1 Pembroke Broadway, Camberley, Surrey, England

Abstract. The Guidelines described are under development within PURDUE EUROPE Technical Committee on Safety and Security (TC7).

The purpose of the Guidelines is to provide a comprehensive reference work to assist various parties, involved in the procurement and operation of safety related computer systems, to determine the scope and contents of associated documentation.

The proposed Guidelines take account of a number of specific influencing factors. The overall structure was determined by consideration of definitively stated principles which reflect the requirements of producers and users of documentation throughout a system's life history. Attention is paid to selecting the optimum documentation for a project.

Although the Guidelines are not yet complete certain parts are well under development. The current status and projected progress are reported.

Keywords. Documentation, Computer System Development, Safety Systems.

BACKGROUND

Motivation for the Guidelines

The Technical Committee on "Safety and Security" (TC7) of the PURDUE EUROPE Workshop consists of members with a diverse background and experience in safety related computer systems. The experience derives from every-day involvement as Users, Suppliers and Researchers.

Early interest within TC7 focused on the need for comprehensive documentation as part of safety related computer system procurement and operation. It soon became clear that there did not exist a single set of standards or guidelines for all the necessary aspects of process computer systems documentation; and certainly none which specifically addressed the needs of safety related computer systems.

Thus, it was decided to form, within TC7, a Documentation Sub-Group with the specific objective of generating a set of Guidelines for all documentation aspects of safety related computer systems. Such Guidelines are intended to assist individual standardisation bodies to achieve a common baseline. The remainder of this Paper relates to work to-date, of the Documentation Sub-Group, towards the objective stated.

Relationship to Existing Standards and Guidelines.

Every endeavour has been made to account for the ideas and requirements of known national and private European and US standards and guidelines. This has been achieved by direct experience of TC7 members and by study of available literature.

Although no formal relationship exists between these Guidelines and any other, implicit or explicit reference has been made to at least those publications known to TC7 members which are listed in the Annex.

OBJECTIVES

Purpose of the Guidelines

The purpose of the Guidelines is to provide a comprehensive reference work to assist various parties, involved in the procurement and operation of safety-related computer systems, to determine the scope and contents of the associated documentation.

Use of the Guidelines by all parties

concerned in a project is intended to contribute towards the production of well structured, integrated and easily useable documentation.

Special Documentation Requirements of Safety Related Computer Systems.

The need for assurance as to the safe operation of safety related computer systems imposes strict requirements on the documentation for design visibility, implementation detail (particularly for software), proof of test cases executed, ease of modification etc.

Documentation meeting these requirements is obviously useful throughout the life of a system from initial conception onwards, and is of particular use to Licensing Authorities.

THE GUIDELINES

Influencing Factors

The Guidelines reflect three major observations regarding safety-related computer systems and their documentation:-

a) A computer system is part of a total system (e.g. including process plant, operators, control systems, environment) and can itself comprise a number of sub-systems.

b) It is necessary for the documentation to permit the total system, or any part of it, to be considered from a number of different viewpoints,

    e.g. operational,
          functional,
          structural,
          locational,
          temporal.

c) The total documentation must cover all aspects from requirement, through design, implementation and operation, together with a number of auxiliary topics; paying special regard to the documentation necessary for safety verification.

Principles Determining the Documentation Set

The documentation set was determined by consideration of the following principles:

a) There shall be an integrated set of documentation as opposed to a loosely-coupled collection of individual documents. Each member of the set must have a defined relationship to other members and contain well bounded subject matter.

b) The documentation of the requirements for the system shall be separate from that of the resultant implementation.

c) The system shall be documented in terms of how it is designed and implemented, as a reference manual independent of documentation relating to the project which led to the implemented system.

d) Documentation concerning the project shall be provided, delineating management matters (standards, etc.) from technical matters (testing, etc.).

e) Specific Maintenance documentation shall be provided.

f) Where proprietary equipment is employed as part of the system or as auxiliary (e.g. test equipment) it too shall be documented (normally by provision of standard manuals).

These considerations led to definition of a documentation set comprising seven parts, as follows:-

1. Documentation Overview and Index.
2. System Requirements.
3. System Description.
4. Technical Support Documentation.
5. Project Management Documentation.
6. Maintenance Documentation.
7. Associated Equipments and Software.

The way in which adherence to these principles leads to the documentation set proposed is depicted in figure 1.

Wherever possible, within each part, the Guidelines require that the documentation be presented in a hierarchic manner. The need to satisfy a number of views (even within one part of the documentation) implies the presence of some overlap: where this occurs special attention is necessary to preserve clarity of information.

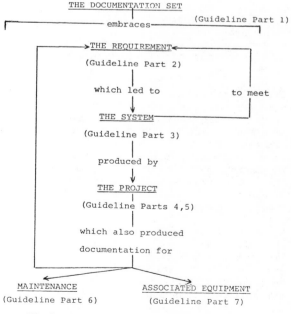

Fig.1. Determination of the documentation set

Scope of Each Part of the Documentation Set

For each part of the documentation set the Guidelines address Subject and Contents in detail and gives some indication of likely Formats, Producers, Users and when produced.

The outline profile for each part is shown in Table 1.

SELECTING THE OPTIMUM DOCUMENTATION FOR A PROJECT

Documentation produced in accordance with the Guidelines is expected to be relevant to the needs of all potential users. However, if due regard is not taken of particular project requirements this documentation may be more extensive and costly than is necessary.

The need to minimise both the cost and the bulk of documentation is well recognised. This can be achieved if a project's specific requirements for documentation are clearly identified.

Towards this end, the Guidelines will identify a number of 'System Characteristics' which, in some degree, are expected to influence selection of the optimum document-ation (in terms of parts required and depth of coverage of each part).

For each characteristic there is an associated five point 'stringency' scale (in general the higher the scale rating, the more strict is the documentation requirment). Thus, a system can be 'scored' for each characteristic and eventually direct guidance will be given at appropriate places in the Guidelines on how the score for particular characteristics influences that particular part.

The characteristics and scales currently being considered are shown in Table 2. It is intended that the characteristics will be sub-divided into safety-related character-istics and more general characteristics.

PRESENT STATE OF DEVELOPMENT

The TC7 Documentation Sub-Group is divided into a number of working parties each producing a draft of a part of the Guidelines. Each such draft is reviewed by the Sub-Group before submission to whole of TC7 approval. These review and approval activities, together with the working party discussions, take place during the four times per year TC7 meetings.

To date, no part has yet been submitted for TC7 approval. However, within the next 12 months at least Parts 2 and 3 (System Requirements and System Description) are expected to have achieved such approval for the first issue.

Production of one section of Part 3 requires specific assistance from another PURDUE EUROPE TC concerned with man-machine interfaces.

Another Sub-Group of TC7 is actively considering Project Management problems and it is expected that they will contribute a draft of Part 5, for review by the Documentation Sub-Group, later this year.

Detailed work on Part 4 "Technical Support Documentation" will commence shortly.

Later it is hoped to publish as appendices, recommendations on alternative formats, symbology, etc., which have been shown by practical use to be of benefit.

ACKNOWLEDGEMENTS

The author would like to record his grateful thanks to all members of TC7 Documentation sub-group, of which he has been chairman for the last two years.

The work of this sub-group has been going on for a number of years and it is a tribute to the members that we are now in a position where the Guidelines have taken firm shape and we can look forward to publication of complete Guidelines over the next two years or so.

ANNEX

Existing Documentation Standards and Guidelines

Unfang der Dokumentation für, Hardware and Software (in preparation); VDI/VDE, West Germany.

Programm Dokumentation, rahmenangaben; DIN 66230.

NCC Data Processing Documentation Standards; The National Computing Centre, 1977, U.K.

Code of Practice for Documentation of Computer Based Systems, BS5515: 1978; British Standards Institute.

Requirements for the Documentation of Software in Military Operational Real-Time Computer Systems (JSP 188); Ministry of Defence, U.K.

COMDOC - A method of specifying a software system by defining the communication between components; Electrical Quality Assurance Directorate, Procurement Executive, Ministry of Defence.

Guidelines for Documentation of Computer Programs and Automated Data Systems (FIPS PUB38); Federal Information Processing Standards Publication, 1974.

Apolio configuration management manual, NHB 8040.2 (formerly NPC 500-1), Exhibit viii: 'Computer Programs'; NASA, Washington, DC., Jan. 1970.

'Guidelines for Documentation of Scientific Software Systems': Trauboth, H. Proc. of 1st IEEE-International Symposium on Reliable

Software, New York, 1973.

'Proposal for hierarchical description of software systems', NASA Technical note TN D-7200; NASA Washington, DC., March 1973.

Development Documentation Systems (DDS) for Weapon Systems and Equipments (NWS 10); Ministry of Defence, 1977, U.K.

TABLE 1 - OUTLINE PROPOSAL FOR EACH PART OF THE DOCUMENTATION SET

TITLE	SUBJECT AND CONTENTS	FORMATS	PRODUCERS	USERS	WHEN PRODUCED

PART 1
OVERVIEW AND INDEX
	Subject: Guide to the total documentation set.
	Contents:
	- Overview of documentation set
	- Index
	- Annex, containing typically,
	  Format/Symbology rules
	  Definitions of Terms
	  Document numbering conventions
		- Text
		- Tables
			- System Supplier	- All parties	Built-up progressively throughout the system life-time

PART 2 SYSTEM REQUIREMENTS	Subject: What the system is required to do, and any associated constraints. Note: This normally comprises information supplied by the customer (e.g. Operational Requirements) and also information generated by the supplier as a pre-cursor to, or as part of, system design. Contents: - Functional Requirements and Facilities - Data Requirements - Communications Requirements - Allowable System States - Modes of operation/Mode Transition (failure defined, operationally defined, etc.). - External Interfaces and Equipment - Specific Features - e.g. Safety, Security, Reliability, Response Times, Legal Requirements, Acceptance Requirements, Expansion Capability, Flexibility, Maintainability. - Installation - Test Strategies	- Text - Tables - Block Diagrams - Flow Diagrams - Formulae - Technical Drawings	- Customer - System supplier	- System Supplier - Licencing Authority - Test Personnel	Requirements Specification Phase. Prior to Design and Implementation Phases.
PART 3 SYSTEM DESCRIPTION	Subject: A comprehensive description of the system under the following headings: System Hardware Software System Operator and User Aspects. Specifically does not make reference to the way the system was developed. Contents: System - Configuration - Principal components/sub systems - Functional specification and description - Interfaces Internal, external - Inter-Component Data transfers, their management and timing	- Text - Tables - Block Diagrams - Flow Diagrams - Timing and sequencing diagrams - Technical Drawings - Circuit Diagrams - Check list - Code listings - Formulae - 'Pictures' e.g. of VDU formats	- System Supplier	- Customer - System users - Licencing Authority	During development implementation, installation

TABLE 1 - (CONTINUED)

TITLE	SUBJECT AND CONTENTS	FORMATS	PRODUCERS	USERS	WHEN PRODUCED
PART 3 SYSTEM DESCRIPTION	Contents continued - Data Location and Management - System Management and Configuration Control Hardware - Detailed hardware configuration including standby features - Technical and facility description at system and sub-system level - Interface data, timing diagrams - Test and maintenance facilities - Installation and environmental specification - Spares, test equipment and consumables - Design, production and test data for bought-in equipment Software - Software specification referenced to Part 2 and the 'system' description (Part 3, Section 1). - Software System Design - Implementation of the Requirements - Components: modules, interfaces - Environment/Support Software - Run-time structure - Data bases - Software/Hardware interface description. System Operation and User Aspects - Start Up/Restart - Shut Down: controlled, emergency - Operating instructions for normal use - Controls and indicators - Operating Instructions				

TABLE 1 - (CONCLUDED)

TITLE	SUBJECT AND CONTENTS	FORMATS	PRODUCERS	USERS	WHEN PRODUCED
PART 4 TECHNICAL SUPPORT DOCUMENTATION	Subject: The way in which the system was developed. Contents: - Test/Acceptance Strategies/Plans - Test/Acceptance Results - Test Discrepancy/Observation Reports - Failure Analysis, MTBF and MTTR Calculations - Change/Modification Records - Studies and Analyses - Certification Report	- Text - Tables - Licencing Certificate - Block Diagrams - Flow Diagrams	- System Supplier - Customer - Licencing Authority	- Customer - Licencing Authority	During development, implementation, installation, operation.
PART 5 PROJECT MANAGEMENT DOCUMENTATION	Subject: The way that the project was managed during development, including management of modifications, changes, etc. The management of maintenance and enhancement of the installed system. Contents: - Project Code of Practice - Project Plan - Project Control Procedures - Change Control Procedures	- Text - Diagrams	- System Supplier - Customer	- Customer - Licencing Authority	Continuously throughout the system life cycle.
PART 6 MAINTENANCE DOCUMENTATION	Subject: How to perform routine maintenance, fault finding, repair. Contents: - Maintenance Procedures - Maintenance Regulations - Diagnostic Charts	- Text - Diagrams	- System Supplier	- Customer - Licencing Authority	Prior to installation
PART 7 ASSOCIATED EQUIPMENTS AND SOFTWARE	Subject: Proprietary equipment and software. Contents: - Test equipment - Program development system - Software library - Test programs - Manufacturers' Handbooks	- Text - Diagrams - Code Listings	- System Supplier	- Customer	Prior to installation

TABLE 2
SYSTEM CHARACTERISTICS

CHARACTERISTIC	SCALE RATING 1	SCALE RATING 2	SCALE RATING 3	SCALE RATING 4	SCALE RATING 5
1. Scope of Safety Considerations	Risk to components or data only	Risk to plant	Health risk	Risk to life of plant personnel	Risk to life of external personnel
2. Scope of Security provisions	Risk of error by trained personnel	Risk of error by untrained personnel	Deliberate error by untrained personnel	Terrorism	Deliberate error by trained personnel
3. Access to System	Only by licensed trained staff	Trained staff, not necessarily familiar with safety regulations	Staff with detail knowledge of the system	Staff generally informed about the system	Others
4. Maximum likely cost of system failure ($)	$\leq 10^4$	$\leq 10^5$	$\leq 10^6$	$\leq 10^7$	$> 10^7$
5. Provisions to limit Spread of failure	Extensive: - redundant components - failsafe components - auto-failure indication - easy replacement	Considerable: Any 3 of 4 from coloumn 1.	Limited: 2 of 4 from coloumn 1	Minimal: 1 of 4 from coloumn 1	Trivial:
6. Reliability required (per year)	$\geq 1 - 10^{-2}$	$\geq 1 - 10^{-3}$	$\geq 1 - 10^{-4}$	$\geq 1 - 10^{-5}$	$\geq 1 - 10^{-6}$
7. Availability required	$\geq 1 - 10^{-2}$	$\geq 1 - 10^{-3}$	$\geq 1 - 10^{-4}$	$\geq 1 - 10^{-5}$	$\geq 1 - 10^{-6}$
8. Component MTBR	0.1 year	2 years	5 years	15 years	15 years
9. Complexity	Single processor, limited multi-programming	Multi-programming	Extensive multi-programming	Multi-computer, highly configured	Multiprocessor network
10. Programming Language	Problem Oriented Language	High Level Language	H.L.L and limited Assembly Language	H.L.L. and extensive Assembly Language	Assembly language
11. Personnel involved in development	Small team: 1 - 2 persons	Medium team: 3 - 5 persons	Large team: 5 - 20 persons	Multiple teams:	Multiple contractors
12. Development cost ($)	20K	20K - 50K	50K - 200K	200K - 500K	500K
13. Span of Operation	Part of a single plant	Single plant	Multi-plant	Company-wide, Country-wide.	Multi-company, International
14. Frequency of change of role	Never	Rarely	Occasionally	Frequently	Continually
15. Originality	None: e.g. re-implementation on different equipment	Minimal: e.g. addition of more stringent performance requirements to existing systems	Limited: e.g. addition of new interfaces	Considerable: e.g. requiring 'state of the art' techniques	Extensive: e.g. requiring advance in 'state of the art'
16. Generality	Highly Restricted: Single purpose	Restricted: Some parameterisation	Limited Flexibility: e.g. Scope for format changes	Multi purpose: e.g. applicable to a range of applications	Very flexible: e.g. applicable to a variety of applications and equipment
17. System Life-time	2 years	2 - 5 years	5 - 10 years	10 - 20 years	20 years

SAFETY CONSIDERATIONS IN PROJECT MANAGEMENT OF COMPUTERIZED AUTOMATION SYSTEMS

H. Trauboth and H. Frey*

Kernforschungszentrum Karlsruhe GmbH, Institut für Datenverarbeitung in der Technik, Postfach 3640, D-7500 Karlsruhe, Federal Republic of Germany
*Technische Dienste und Qualitätssicherung, Geschäftsbereich Elektronik, BBC AG, Brown Boveri & Cie, Ch-5401, Baden, Switzerland

Abstract. A number of safety measures and analyses have to be considered during the entire life cycle of a computerized automation system. The implementation and effectiveness of these safety measures have to be assessed and controlled by proper safety management procedures. The safety management organization must guarantee independency of safety control.
A framework of guidelines for safety management of computerized automation systems is presented.

Keywords. Safety management; process control computers; project management; safety; reliability; system analysis; configuration management.

INTRODUCTION

Safety measures in project management have been developed and continuously been improved for a long time for the development and operation of technical systems which affect the safety of human life and property, but do not contain computer control.

A special new type of management has emerged which deals particularly with safety considerations in project management; it is called safety management and is now well established in safety-oriented technologies such as nuclear technology, aviation, railway and military systems /1-4/. A number of procedures, guidelines and standards have been developed by governmental authorities, technical committees and professional organizations. However, computer systems have mostly been excluded. Computers are used mainly in a passive role such as for data logging in highly safety-oriented nuclear systems. In aviation and military systems, they play already a more active role such as for flight and landing control. Nevertheless, safety management for the development and operation of computer systems is not well established yet.

The subgroup 'Project Management' of the Technical Committee 7 (TC 7) 'Safety and Security' of the European Purdue Workshop began to develop a framework of guidelines for safety management of industrial process-control computers. This paper presents the results of discussions within this subgroup and additional ideas of the authors on the structure and contents of such guidelines.

Existing safety management guidelines for technical systems cannot be used directly since computer systems for industrial applications have special properties.

SPECIAL PROPERTIES OF COMPUTER SYSTEMS FOR INDUSTRIAL APPLICATIONS

The number of components and their interactions of a computer system is an order of magnitude larger and more complex than that of other technical systems. Since the development of microelectronic technology processes rapidly, a generation of computer models changes about every five years. Exhaustive testing and maintenance of such systems is not possible because of prohibitive time and cost.

The software, which makes the computer system run, consists usually of ten-thousands of instructions acting like a complex interrelated network. While hardware errors during operation may be caused by physical wear or aging, software errors are actually design errors. Since the number of possible errors and their effects is tremendously large, failure analysis methods commonly known often cannot effectively be applied. Also, software undergoes frequent changes which complicates maintenance. Statistics on software errors during operation and reliability figures on software practically do not exist. Assessing the consequence of a hardware or software error on the safety of the plant controlled by the computer is often very difficult and calls for over-proportional expenditures.

Therefore, high reliability of the computer system is a precondition for safety. A process-control computer system being integrated into and interacting with its environment (technical plant, process) in real-time can affect the safety of that environment in many ways. This situation requires specific constructive, fault-tolerant and analytical measures for the computer system to achieve the safety of the environment. The application of these specific safety measures plays an important role in safety management of process-control computer systems.

DEFINITION OF SAFETY OF A PROCESS-CONTROL COMPUTER SYSTEM

A process-control system is embedded in a total technical system which we simply call 'system' (Fig. 1).

The computer system receives signals directly from the plant it controls or monitors, or indirectly via a human operator. It sends signals directly to the plant or indirectly via the human operator.

We can define three zones of the plant which are differently affected by the computer system (Fig. 1):

- zone A which is not affected at all;
- zone B which is affected, but errors in the computer system do not lead to hazardous situations;
- zone C which is affected and errors in the computer system can lead to hazardous situations, i.e. the safety of the total system can be endangered.

In safety management, we are concerned with the affects of the computer system on zone C only. For this reason, it is one of the first tasks in the requirements analysis phase to determine the boundaries of zone C by a preliminary hazard analysis and failure cause-consequence analysis.

In the hazard analysis, dangerous states of the plant and events that can cause them are identified. The cause-consequence analysis determines those hazardous events which could be caused by the computer system.

In hazard analysis, it is assumed that a hazardous event in the computer system (incl. interfaces) affects the process or plant that is controlled or monitored by the computer system but does not affect the computer system itself. For instance, an erroneous signal from the computer to a pressure valve may cause a vessel to burst and injure personnel working in the plant. It is not assumed that any erroneous signal could be generated by the computer that would cause a hazardous situation in the computer itself. In a general sense, a hazardous event in the computer system affects the safety of the total system, e.g. in an aircraft, the flight computer may cause a damage of the engine which may lead to a crash or destruction of the plane and the death of the pilot, i.e. destruction of the total system. The computer could cause a hazardous situation also in monitoring if the operating personnel receives such erroneous information that it causes a hazardous situation in the process or plant, e.g. by turning the wrong manually operated valve. Therefore, when we speak of the safety of the computer system (hardware and/or software) we mean actually the safety of that part of the plant which is affected by the actions of the computer system /14/, /16/.

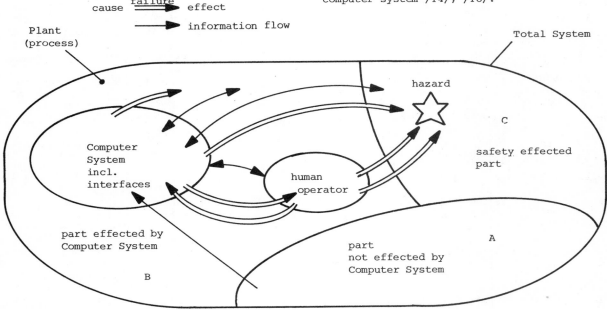

Fig. 1. Failure effects of process-control computer system

STRUCTURE OF PROPOSED GUIDELINES

Objectives of Safety Management for Computerized Automation Systems

Safety management aims at assuring that the computer system inherently does not send erroneous signals to the plant being controlled by the computer and does not present erroneous information to the human operator of the plant such that the plant is never placed into an unsafe state. Safety management has to ensure that the optimal constructive measures in the design, proper quality control, sufficient testing and procedures for the operation of the computer system under severe environmental conditions are employed within a reasonable cost-benefit range to obtain safe operations meeting standards and legal regulations /4/.

The development of safety built into a process-control computer system and the assurance of safety by quality and safety control is an ongoing process during the total life cycle of a system. Therefore, the life cycle serves as a time base for the safety management approach.

During each phase of the life cycle, safety-relevant actions have to be performed by development and by operations personnel on one side and by quality/safety-control personnel on the other side. The development and operations personnel perform actions which primarily result in constructive measures, while the quality/safety-control personnel perform actions which analyze, evaluate and verify the performance and quality of these constructive measures.

Safety management deals with reviewing and controlling these constructive and analytical measures. To guarantee independence of safety management and control, a special safety management organization is to be set up.

Safety-relevant Actions

There are a number of different actions which have to be conducted during the life cycle of a system to assure a system's safety. In a safety management guidelines handbook, not all actions can be described because there are too many of them and they are dependent on the particular application. In this paper, we try to categorize safety-relevant actions and measures for each phase.

These actions are presented in form of
- check-lists,
- with priorities (importance) attached,
- sequences of actions and timing if necessary,
- results (products) and documation of actions,
- examples for explanation and illustration.

These actions may be constrained by economics such as costs and by legal regulations.

We distinguish between two types of actions, i.e.:
- constructive and
- analytical measures.

The constructive measures are
- technical schemes and
- operational schemes
- supported by tools

to achieve
- reliability and safety,
- fault-tolerance,
- testability and maintainability,
- transparency of system structure

built into the design and implementation of the computer system.

The analytical measures are
- analysis
- test & validation
- supported by tools

to assure that the products (software, hardware) meet the criteria establish for
- reliability and safety,
- fault-tolerance,
- testability and maintainability,
- transparency of system structure.

The criteria are established at the beginning of the development process.

Safety Management Control

Safety management control should assure that all necessary safety-related actions of each phase have been performed satisfactorily. It consists of
- check-lists of audits and reviews of products, operations and documentation with regard to reliability and safety,
- decisions to be made,
- control actions concerning proceedings with development and operation,
- procedures and documentation for management control,
- organizational units involved,
- control of subcontractors.

Management Organization

There are three major organizational areas to be considered:

- the line functional units which develop the computerized automation system, those which operate the total system and those which perform quality assurance;
- the project management units which coordinate the development, review the products and the process of the development. The criteria for their reviews are performance, cost and quality of the products;
- the safety management unit, which reviews and audits the products and progress of the development and operations with regard to safety. Large institutions may be able to afford a safety management unit separate from the quality control unit.

These organizational areas have different responsibilities and authorities. The information flow and decision flow within and between them must be well established.

The management organization is considered to be constant for all life cycle phases.

SAFETY MANAGEMENT DURING LIFE CYCLE

According to the structure outlined in the previous chapter, the safety-relevant actions and safety management controls are briefly described for each phase of the life cycle. They are somewhat representative but not comprehensive and they reflect the discussions of the subgroup 'Project Management' of TC 7, European Purdue Workshop.

Requirements Analysis Phase

Safety-relevant actions.

1. System requirements review. Study and review all functional requirements of the computerized automation system and identify those requirements which effect the safety of the total system directly or indirectly. Determine the boundaries of these effects with support of the Preliminary Hazard Analysis (PHA).

2. Preliminary Hazard Analysis (PHA). A preliminary hazard analysis shall be performed as an integral part of the system requirements review to identify potential hazard and inherent risks associated with each subsystem of the plant that the computer system controls and/or monitors.
The preliminary hazard analysis summary report should

a) identify potential hazards and methods planned to eliminate or control them;
   Examples:
   - Critical failure modes and states of process equipment to be controlled/monitored;
   - Safety devices and safety configurations in the plant (e.g. interlocking controls, redundant parts, back-up components);

b) outline undefined areas requiring guidance or decisions.

c) Perform simulations of plant for failure-consequence analysis.

3. Regulations. Determine existing and anticipated standards and regulations (legal aspects) for safety and environment, which will influence the design of the computer system (hardware and software) /4/.

4. Preliminary safety study and analysis.

a) Study failure statistics and failure-countermeasures of similar systems with the aim to utilize them for the development phase.

b) Highlight special areas of safety considerations such as system limitations, technology risks and man-rating requirements which constrain the computer system design.

c) Determine additional functions to be incorporated into the computer system to meet all safety requirements of the total system.
(E.g., in a chemical plant, leakages of gases which may cause environmental damage should be detected, recorded and documented by the computer system via additional instrumentation. Or, in a transportation system such as BART, which is installed in an earth-quakeprone area, the computer system may monitor tremors along the roadbed in order to signal the trains to reduce their speed or to stop in order to prevent an unsafe operation).

d) Identify possible safety interfaces problems concerning interfaces between process equipment/operators and computer system.

e) Determine skill and background of operational personnel and estimate required training effort in operation of safety-oriented equipment, incl. continuous training for handling critical failure situations.

f) For evaluation of economic constraints estimate and specify the costs allowed for the development and operations of the system as they constrain the total development costs and extra features for safety-oriented design.

5. System safety requirements.

a) In utilizing the results of the preliminary studies and analyses of 1. to 4., list the global safety requirements (e.g. the state and time the total system has to transit to if a critical failure occurs).

b) Derive a system concept and the detailed system safety requirements, i.e. technical functions, their performance characteristics, special system structures and operational functions, that are particularly needed to meet the global safety requirements.
Identify those which are demanded by law and regulations.

Example: A global safety requirement may be the following:
If a brake failure in a train occurs, the speed of the train (state of system) has to be reduced within 30 sec (time for transition) to 50 % of the normal speed (failure-mode state of system) without increasing risk of accident.

Detail Design and Programming Phase

Safety-relevant actions

1. *Safety design specifications.* Determine the safety design specifications to be incorporated in the equipment and software detail design based on the system safety design criteria. In compiling the safety design criteria, all events that can contribute to creating the hazard identified in the PHA should be considered. Document all safety decisions and maintain a current, accurate system safety file.
(E.g. in the detail design of hardware, comparator circuits which detect hardware errors and switch redundant components in case of an error are considered.
E.g. in the detail design of software, a program module which controls a critical valve in a plant is allowed to act only after being released by another program module or hardware component that interlocks in case of a logical error or failure of a plausibility check). Use diverse implementations for the same critical function /5/.

2. *Programming rules.* Establish programming rules for reliable structuring and implementation of the software design into programs /6,7/.
Critical parts should be programmed by independent programming teams using different programming languages and algorithms for realizing the same safety-oriented functions /8/.

3. *Programming.* During this phase, safety-oriented software modules as specified during the Software Detail Design are being implemented. The programs should be implemented according to the reliable programming rules as defined above.

4. *System safety hazard analysis (SSHA).* Update and expend the PHA into a complete System Safety Hazard Analysis (SSHA) for computer hardware and software. This analysis should be complete by the final design review.

5. *Safety and quality control.* Adequate procedures shall be invoked through the planned, controlled and scheduled system quality control to insure that safety achieved in design is maintained and verified during programming. Corrective action shall be taken to eliminate, reduce and control hazards so identified. These corrections shall include necessary changes to software documentation. An audit shall be performed to identify any new system safety hazard which may result from the introduction of engineering changes. The impact of such changes on safety shall be evaluated to determine whether the previously established safety level of the system has been maintained; if not, redesign or change procedures shall be initiated to obtain the contracted level of safety.

Use of tools

A number of automated test systems are available which can be applied during this phase.

Documentation

Various kinds of representation of the detail design and programming can be used such as Structured Charts, decision tables, flow charts and structures commentaries in program listings. They are detailed in the 'Guidelines for Documentation' prepared by the TC-7-subgroup 'Documentation'.

Safety management control

1. *Reviews and audit.* Perform detail design and program reviews using walk-throughs of documentation and listings to assess the status of compliance with the safety program objectives and software design/programming rules. This review shall identify any deficiencies of the system with respect to safety and provide guidance for further development if necessary.

2. *Safety test and verification.* During this phase, the implementation of the safety design requirements and specification should be checked and verified by appropriate tests and analyses. Unsafe conditions should be simulated to test the performance of corrective actions and to test the theoretical assumptions about primary and secondary failure effects during the hazard analyses.
Various kinds of tests (software test, system test) are to be performed using software models and real-time simulators for simulating the actual normal and abnormal plant signals interfacing with the computer /9,15/.

3. *Decisions and procedures.* List decisions to be made and procedures used in executing safety management.

Operations Phase (incl. Maintenance)

Safety-relevant actions

1. *Safety data collection.* Collect data from system deficiency and analysis reports submitted by operating and maintenance personnel.
The hazard and failure data are maintained in the Safety Information and Documentation File, which can be utilized to improve the safety design and avoid unsafe conditions in future products which may occur in the system and detail design phases of development.
Control and document any implementation and design changes and the results of hazard analyses relevant to these changes.
This file should contain an accurate account of all analyses and corrective actions taken by the system safety office and operations manager. The data should be stored in such

a format that easy and cost-effective updating and retrieval of the file is possible. This file should serve as a data base for safety reports and for utilization during the design of safety-oriented systems.

For the development of safety-oriented computer system (hardware, software), safety data and reliability data of hardware devices (of similar existing process equipment, computer hardware) should be used if appropriate. These data should be furnished in proper format to the developing organization. The contractor shall maintain liaison with other data sources to enable identification and evaluation of hazard and safety design deficiencies.

2. Analysis of failures. Analyze system failures which caused or could have caused unsafe states and initiate corrective action.

Documentation

Operational handbooks are used by the operators which contain safety procedures to be followed in case of system failures.

Failures and their consequences are documented as much as possible by a computerized Safety Information and Documentation File System (see paragraph 'Safety Data Collection').

Safety management control

Audit operational safety of system to determine if design, operating and maintenance procedures, and emergency procedures are adequate based on user experience. Evaluate design changes and modifications to computer system (hardware, software) to ensure that safety is not degraded. Review continually operator and maintenance procedures and documentations to ensure that safety requirements, procedures and cautions are adequate.

SAFETY MANAGEMENT ORGANIZATION

In project management of the total system, the project manager responsible for the computer system plays a special role since the computer system interacts closely with all other subsystems of the total system as is indicated in fig. 2.

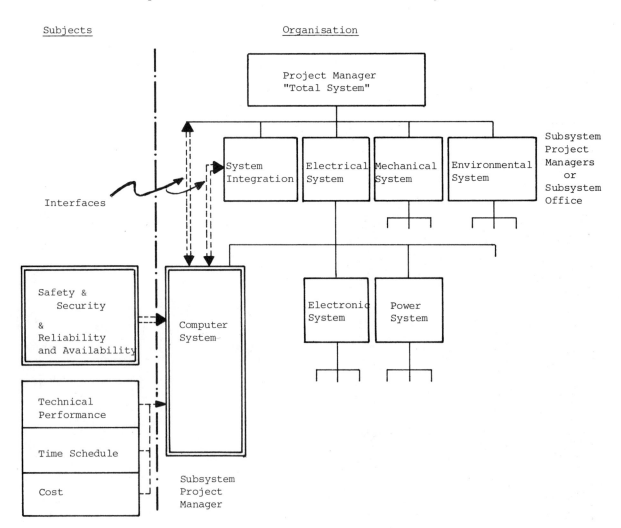

Fig. 2. Project management of a total system (overview)

Although the computer system is an electronic subsystem by its physical nature, it should be treated as a special subsystem from the managerial point of view on the level of the System Integrations Office (or Subsystem Project Manager). For this reason, the safety office should also contain a special unit handling exclusively safety problems of computer systems and being responsible for controlling the implementation of the guideline on computer system safety requirements.

The responsibilities and functions of those directly associated with system safety policies and its implementation should be clearly defined. The authorities delegated and the relationship between line, staff, interdepartmental and project organizations and their management shall be identified.

A clear distinction should be made between personnel responsible for monitoring and reviewing the safety requirements of the design and of the implementation and the personnel responsible for the design and implementation. Probably, the quality control personnel might be best suited for safety control.

CONCLUSION

The safety considerations in project management and first proposed guidelines for safety management of computerized automation systems are not complete yet. They have been derived in discussions held at several meetings of TC 7, European Purdue Workshop /10-13/. Guidelines on documentation, software and hardware construction rules, testing tools and terms, being derived in TC 7 may finally become referenced parts of a manual on project management requirements for safety-oriented computerized automation systems.

The authors wish to thank their colleagues who participated in the discussions on project management Mr. Aitken, Dr. Genser, Mr. Hendry, Mr. Levene, Mr. Reinshagen and Mr. Winter for their contributions.

APPENDIX

Safety Analysis Methods

During the entire life cycle, a number of analyses should be performed to identify hazardous conditions and to eliminate and control them. They also should determine, evaluate and up-date appropriate safety requirements for design, test and operations. Quantitative analysis to assess numerically the relative safety of a computer-controlled system is presently still not possible. However, qualitative analysis can evaluate the relative safety of alternative designs and trade-offs.

Particularly the summaries of the analyses should be described in such a way that they are transparent to legal personnel if systems of public interest are involved.

The <u>hazard analyses</u> deal with those areas of the plant or process which are controlled and/or monitored by the computer system. Hazards in the plant may be caused by unsafe conditions in plant or process areas which
(a) are out of the region of computer control;
(b) are within the region of computer control and are due to
  • computer system design (hardware, software) oversight,
  • failures of the computer system.

The hazard analysis determines the relationships for each hazard between cause, effect, severity of effect and corrective action or minimizing provisions during each phase of the development process. The probability with which the hazard may occur and the costs of the corrective action should be determined too. The hazard analysis may be conducted by less formal methods such as hazard cards, event expansion or formal methods such as Boolean logic or fault tree analysis. Simulations should be applied whenever appropriate. Hazard analysis at the system design level identifies coarsely possible hazards caused by subsystems, software modules and interfaces. At the detail design level, smaller modules are analyzed /3/.

The functions of each computer system software and hardware module should be analyzed to determine if any design error in the software or any performance degradation or functional failure in the hardware could result in hazardous conditions. The analysis should include the determination of failure modes and delineate safe operating states to which the system can revert in case of failures. Safety margins (mostly needed to compensate for uncertainties in system implementation) and tolerable levels of degradation should be determined by theoretical and simulation approaches. Criteria and methods for judging safety levels (including measurable variables, measurement methods and operational procedures) should be established. The completeness and consistency of the safety requirements should be demonstrated if possible. Hazards caused by human operator intervention via computer-controlled consoles especially during abnormal operational conditions should be analyzed. Procedures for manual use of such consoles should be evaluated and changed if necessary.

A (qualitative) <u>risk analysis</u> should be conducted which estimates the risks involved in using advanced technologies as opposed to well-proven technologies. This applies particularly to new microelectronic devices such as microcomputers and their interfaces in rough environment.

REFERENCES

/1/ MIL-STD-882 (1969). Military Standard, System Safety Program for Systems and Associated Subsystems and Equipment: Requirements for US Dept. of Defense (DOD).

/2/ Rodgers, W. P. (1971). Introduction to System Safety Engineering. John Wiley, New York.

/3/ Brown (1976). Safety Systems Analysis. Prentice Hall, International.

/4/ IEEE Standard for Qualifying Class IE (1974). Equipment for Nuclear Power Generating Stations. IEEE Std. 323.

/5/ Gmeiner, L., and U. Voges (1979). Software Diversity in Reactor Protection Systems. IFAC-Workshop SAFECOMP '79. Stuttgart.

/6/ Voges, U., and W. Ehrenberger (1975). Vorschläge zu Programmierrichtlinien für ein Reaktorschutzsystem. KfK-Ext. 13/75-2, Kernforschungszentrum Karlsruhe.

/7/ Ehrenberger, W., and J. R. Taylor (1977). Recommendations for the Design and Construction of Safety Related User Programs. Regelungstechnik, 25, 2, 46-53.

/8/ Gmeiner, L., and U. Voges (1979). Software Diversity in Reactor Protection Systems: An Experiment. IFAC-Workshop SAFECOMP '79. Stuttgart.

/9/ Voges, U. (1976). Aspects of Design, Test and Validation of the Software for a Computerized Reactor Protection System. 2nd Intern. Conference on Software Engineering, San Francisco.

/10/ Trauboth, H. (1975). Methods to develop Safe Computer Systems. Purdue Working Paper No. 8 and 35.

/11/ Trauboth, H. (1977). Proposal for Modification of Military Standard MIL-STD-882. Requirements for System Safety Program for Systems and Associated Subsystems and Equipment. Purdue Working Paper No.129.

/12/ Trauboth, H. (1978). Proposed Guidelines for Safety Management of the Development Phases 'Requirements Analysis' and 'Functional Design'. Purdue Working Paper No. 152.

/13/ Genser, R. (1976). Project Management Patterns for Safety Related Systems. Working Paper TC SS No. 66. European Workshop TC 7, Safety and Security.

/14/ Frey, H. (1979). Safety and Reliability - Their Terms and Models of Complex Systems. IFAC-Workshop SAFECOMP '79. Stuttgart.

/15/ Voges, U., and J. R. Taylor (1979). A Survey of Methods for the Validation of Safety Related Software. IFAC-Workshop SAFECOMP '79. Stuttgart.

/16/ Frey, H. (1976). A General Safety Model and the Relation to Reliability and Availability. Purdue Working Paper No.45.

DISCUSSION to the papers by A.A.Levene and H. Trauboth/H. Frey

Lauber: A very important problem is, how to make the documents so understandable, that people really understand what is written there.

Levene: The documentation is a system in itself. If you want to ensure quality of the documentation, you must apply equivalent measures to those you use to ensure the quality of the software system. No documentation should be released until it has actually been "used".

Lauber: This makes documentation even more expensive.

Trauboth: Documentation should be considered the essential product of the software development; software itself mainly consists of documentation. It is difficult to obtain agreement among people how to write documentation, so that different professionals such as designers, implementers and users can read and understand it.

STANDARDS FOR THE PRODUCTION OF HIGH QUALITY SYSTEMS

G. P. Mullery

Systems Designers Limited, 1 Pembroke Broadway, Camberley, Surrey, England

ABSTRACT

This paper describes the underlying principles and structure of a set of standards for the production of high quality software. The standards were produced under contract to the European Space Agency. The current status of the standards and proposed future work are outlined. The long term aim is the integration of enhanced aids and well-defined production methods.
Keywords: Software Development Standards.

INTRODUCTION

This paper describes the underlying principles and structure of a set of standards for production of high quality software. The standards described were produced under contract to the European Space Agency (ESA) by Systems Designers Limited.

An initial study phase for the contract considered the requirements for standards and their relationship with future ESA fault tolerance requirements for on-board satellite systems. A second, definition, phase defined the detailed standards, which are in process of publication as a set of manuals.

The delivered standards must be considered as experimental since they have not been used on any significant project.

Both during the currency of the study and definition phases of the ESA contract and since, Systems Designers Limited have been contracted by other organisations to produce software standards for a number of different applications and different development/procurement environments. As a result of these experiences it is now believed that a generalised set of standards can be produced that can be readily adapted to a wide range of project situations.

The generalised set of standards will include a number of refinements, mainly arising from in-house experiment. Further work in this area by Systems Designers Limited will include preparation of associated training material and the integration of standards, methods and automated aids in a comprehensive range of products for use in the development and maintenance of high quality software.

STUDY PHASE SUMMARY

The study phase resulted in four reports (refs. 1,2,3,4). One (ref.4) was produced under sub-contract by the Danish RISØ organisation. The main conclusions are briefly described below.

Purpose of the Standards

The standards were required to aid the production of ESA satellite on-board software systems. The characteristics required of such software are that it should be:

- reliable and demonstrably reliable before flight
- robust
- correct
- easily modified (i.e. maintainable in flight)
- well documented
- constructed from component parts that are independent through modularity of function and implementation.

Use of standards cannot guarantee the required characteristics. They can only enhance the probability of their achievement.

The standards are aimed to be manually applicable, but compatible with the use of automated aids, if they become available.

The standards are aimed at projects with a high reliability requirement. It was intended that they should be relatively easy to tailor for less exacting project requirements (e.g. where there is a low cost of and easy access to perform repairs).

Finally, ESA required that it should be possible for staff with relatively little software experience to check for adherence to the standards. The concept of testing end products of all project phases was thus divided into two areas: Standards Monitoring and Technical Monitoring.

Standards Monitoring tests are exact test definitions whose application requires no technical judgement. Technical Monitoring tests are basically qualitative, requiring the application of technical judgement. The latter are not strictly tests for adherence to standards. They refer more to adherence to a code of practice.

What Standards Are

Standards embrace the definition of both procedures and end products for all necessary project functions. For long-term effectiveness, standards are defined for the entire life-cycle of a system, from Project Initiation through to the end of its Service Life.

In general, standards relate to development tasks, where development is taken to include both management and production activities. Development standards are associated with quality control tests and control of the storage and release of technical products. They include procedures to ensure that products are submitted for test, tested and securely stored.

In a wider sense, there is a need for standards for checking the quality of the standards themselves, based on experience with their use.

The standards produced under the contract with ESA addresses only a subset of the full range of possible standards. They applied to software end-products and their interface with management of software production.

Where Standards Normally Fail

There have been a number of industry attempts to produce standards for the production of software. They have normally failed for one or more of the following reasons:

- they covered only part of the software life cycle
- they were inapplicable (e.g. even for the phases covered, necessary information was omitted)
- they were wrong (e.g. procedures or standards were incompatible)
- they were unclear (e.g. end-product requirements were ambiguously defined)
- they were impractical (e.g their use caused unnecessary delays and difficulties)
- they were over restrictive (e.g. they imposed constraints which applied only to particular types of project)
- there was no genuine management support for their use (e.g. pressure was put on production staff to get end-products out on time, whether they adhered to standards or not)
- there was no project staff motivation for their use
- there was no feedback of information to enable the other problems to be proved and removed.

Some Basic Principles

The most important basic principle is the realisation that any set of standards may suffer from any, or all, of the problems identified above. The first version of the ESA standards must be regarded as experimental. They are quite likely to exhibit some of the problems. Extra care will be necessary on a first application.

To allow a 'safety valve' for standards problems, a mechanism should be provided for approval of standards infringements. Other principles, aimed at removing or reducing the tractable problems with standards, are:

- the standards should be organised so that users do not have to search through large manuals to find the standards which apply to their current task
- the standards should be organised so that experienced users need only checklists to provide quick reminders of what is required: inexperienced users should be provided with a separate commentary which explains the standard (reasons for it, examples etc.)
- the standards should be organised so that removal, replacement or change of existing material can be quickly achieved after inadequacies have been detected
- the standards should clearly indicate to users that any other necessary information, not explicitly required by the standards, should still be supplied
- the standards should require the collection of error reports, problem reports and change requirements, so that there is reliable feedback of information on use of standards, methods and tools available to project staff
- there should be training for all users of the standards before exposure on a practical project.

The ESA contract did not cover the provision of training material, but some independent effort has since been devoted to the problem by Systems Designers Limited.

Standards and Fault Tolerance

Software problems remain after even the most thorough testing practicable for a project. They arise because of:

- hardware failure
- implementation errors (software or hardware)
- input data errors
- data errors resulting from earlier errors

- errors in the system specification.

Fault tolerance techniques aim to reduce the impact of the residual problems. Some future space applications may be expected to require practicable fault tolerance.

Study of fault tolerance indicated four types of system requirement:

- high availability
- no loss of input
- freedom from output error
- continuity of operation.

Not all these requirements apply to every fault tolerant system.

A separate study report (ref.4) defined the basic techniques available for achievement of fault tolerance. The report indicates which techniques apply to which fault tolerance requirement and outlines a procedure for determining the techniques applicable to a given system.

The techniques were considered for inclusion in standards. For standards limited to checks performed manually on software end-products it was judged impractical to make them part of the standard.

The reasons for the exclusion of techniques were normally one or more of the following:

- they were not software techniques
- they required automated assistance for practical implementation or enforcement
- they required qualitative assessment by skilled software producers for enforcement
- they represented a strategy, rather than a definition of specific end-products with uniquely definable contents.

Provision of fault tolerance was thus seen as part of the project code of practice, for the purposes of the ESA standards.

With automated assistance, some fault tolerance techniques could be included in the standards. For example, it would be possible to require (and, if high order languages are used, automatically implement):

- array bound checks
- stack overflow checks
- data assertion checks
- base and limit registers
- trial or reference input checks

In the longer term, it may be possible to define widely acceptable standards for strategies such as:

- redundant programming
- re initialisation
- state re-establishment

THE STANDARDS

Project Functions and Team Structure

The standards recognise four basic types of project function: management, production, quality control and configuration control. Management and production functions are referred to collectively as development.

Standards relating to the management function cover only the interfaces between management and the other project functions (e.g. plans, task definitions, reporting requirements, progress reports).

For the quality control standards, allowance has been made for monitoring adherence to standards (standards monitoring) and code of practice (technical monitoring). The principal aim, for quality control, was to produce manually applicable standards monitoring tests for each defined production and management end-product.

The standards do not enforce a particular division of responsibilities among project staff. Hence, for example, one person may have both management and production responsibilities.

For maximum quality it is recommended that one person should have only one type of responsibility at any one time. This has cost and effort implications for small projects. Whether the policy should be implemented for a small project depends on factors such as the cost of failure and predicted length of software service life.

Whatever happens, if high quality is required and only manual checking is available, it is very strongly recommended that no person should be responsible for approving his/her own end-products.

The standards aim to ensure that the required information is produced, whatever the team structure. They define the required flow of information between project functions as shown in Figure 1. If one person is performing management and production tasks, he/she must produce a task definition (management end product) for the production tasks performed.

As can be seen from Figure 1, configuration control is the hub of all official information flows for the project.

The Project Life Cycle

Each phase produces characteristic end-products. Different parts of a system may theoretically be in different phases of production, but it is desirable to control production, so that there is no major loss of synchronisation.

To permit synchronisation without being over-restrictive, the project life cycle standards were divided into five phases (Figure 2). Within each phase there are standards for development (management and production), quality control and configuration control.

A major review at the planned end of each phase decides whether to continue to the next phase, rework some of the current phase, or cancel the project. Typical of such a review in conventional projects would be assessment of the result of acceptance tests at the end of implementation, and before entering service life.

Standards Structure

The standards must be viewed in two ways: how they are to be stored and how they are to be used. How they are used is dealt with under a separate heading.

For the purposes of storage, the standards can be kept in manuals, like most 'conventional' standards. There is considerable flexibility in how they may be split. One method is shown in Figure 4. This is not identical to the method chosen by ESA which does not separate standards and commentary volumes and has guidance notes combined with the general volume.

The guidance notes consist of narrative overviews of the standards as a whole and of each individual phase. The notes include leaflets which show where to find standards and procedures for particular project functions.

The general standards cover procedures and end-products which apply to every project phase (e.g. procedures for job submission, error reporting, standards for progress reports, minutes).

There are five phase-specific volumes. Each volume contains standards and procedures which are characteristic of only that phase. For example, the requirements specification phase contains standards for technical definition of messages (input/output), processes, system data (internal data), hardware types, etc.

For each standards volume there is a corresponding commentary volume which provides additional explanation to assist inexperienced users.

Each standard volume (and the corresponding commentary volume) is divided into three parts: development leaflets, quality control leaflets and configuration control leaflets.

The lowest level of subdivisions form the basis of a policy for practical use of the standards, as explained below.

Use of the Standards

Practical use is usually inhibited by handing out large manuals. Many people will be aware of examples of standards or procedures manuals which sit in a corner of each office to which they are issued and gather dust. Yet, frequently, they contain quite practical information.

The reason for their neglect is as least partially because the potential user has to sort through quite large quantities of materials which, while relevant to other tasks, is not relevant to the task in hand.

If issued as a total standards manual, the same fate would befall the ESA standards. The wide scope of the standards might make it even more of a problem with the ESA standards.

The recommended method of use is as follows:

- there is one master set of manuals which resides with one person, who has responsibility for standards management
- the standards manager should issue the standards, only when their use becomes appropriate
- only parts of the standards should be issued to any one member of project staff (those parts which relate to their current responsibilities).
- when the parts issued are no longer in active use (e.g. moving to the next phase, or change of responsibilities), they should be returned to the standards manager.

In this way, for example, development staff in a particular phase need only two sets of leaflets:

- development leaflets from the general volume
- development leaflets from the appropriate phase specific volume.

The result is that for current responsibilities, project staff see a quite manageable set of leaflets.

If staff have, as recommended below, been properly trained, only the standards need be issued in this way. The commentaries can be stored as a set of volumes whose normal fate is to gather dust, or act as a door-stop.

Application of the Standards

The standards, as they currently exist, must be viewed as an experiment. They may suffer from any or all of the problems attributed to other standards. The main aim in use must be to progressively remove those problems.

Systems Designers Limited have performed small experiments with use of tailored subsets of the full standards. These have led to knowledge of some problems, for which solutions have been defined.

The first application should preferably be on a pilot project whose major aim is to detect and remove problems.

The application should be monitored on a weekly basis at the start of each phase. The highest priority should be given to the removal of problems. Project staff should be aware of the purpose of the exercise and the project leader should approve standards infringements resulting from known errors in the standards.
All project staff should be trained in use of the standards before they need to apply them. Staff training should, for all project staff:

- explain the basic principals
- identify the types of project function
- demonstrate the structure of the standards
- show how the standards should be used

For staff performing a specific project function (management or production or quality control or configuration control), the training should:

- explain the specific function and its relationship to other relevant functions
- explain the meaning of the terminology used
- give practical examples of use of the standards
- make small experimental use of the procedure associated with example end-products.

FUTURE DEVELOPMENTS

As mentioned in the Introduction, a set of standards embracing refinements is in process of production. The most immediately obvious change compared to the ESA standards is a significant reduction in size, due largely to more efficient use of general standards.

For any serious intending user the ability to provide suitable training, as in the section on Application of the Standards, is vital. To this end, production of training material is currently in progress.

On a wider view, we see standards as only one part of a three-pronged approach to the development of high quality software systems. We see a strong need for automated aids and formal methods of production.

We intend eventually to provide an integrated range of products in all three areas and experiments are already in progress on all three fronts.

We have defined a method, applicable to requirement specifications and design expression and are in process of a co-operative application with a customer who has the automated aid, PSL/PLA(ref.5) as a support tool.

We are in process of updating one our earlier, less expensive, standards installations for another customer, who is also considering the use of PSL/PSA as an aid to a major project.

Finally, we are looking at the need for a mini/microcomputer-based automated aid for system description. We have identified the major functional requirements of such an aid and are in process of determining the level of commercial demand for such an aid.

ACKNOWLEDGEMENTS

The ESA Contract (No. 3385) referred to in this Paper provided a major opportunity to bring together a lot of ideas based on the experience of the author and his co-workers at Systems Designers Limited.

The forbearance of the ESA staff at ESTEC, responsible for monitoring this contract and for approving the results, is acknowledged.

Systems Designers Limited record their appreciation of ESA's permission to freely make reference to the work undertaken on the contract.

REFERENCES

1. SR1, May 1978
   Section 1 - General Introduction and Overview.
   By G.P. Mullery (Systems Designers Limited).

2. SR2, April 1978
   Part 1 - Definition of the Requirements for Standards Relating to Software Development for Spacecraft On-Board Software
   By G.P. Mullery, M.G. Bailey, M.P. Rathbone (Systems Designers Limited).

3. SR3, May 1978
   Part 2 - Feasibility of Storing Specification and Design Information on a Database.
   By G.P. Mullery, M.G. Bailey, M.P. Rathbone (Systems Designers Limited).

4. SR4, May 1978
   Part 3 - Rationale for Selection of Fault Tolerance Techniques for On-Board Software and Options Available
   By J.R. Taylor (RISØ, Denmark).

5. IEEE Transactions on Software Engineering, Vol. SE-3, No. 1, Jan 1977, pp 41-48 PSL/PSA:
   A Computer-Aided Technique for Structured Documentation and Analysis of Information Processing Systems
   By D. Teichrorw and E.A. Hershey III

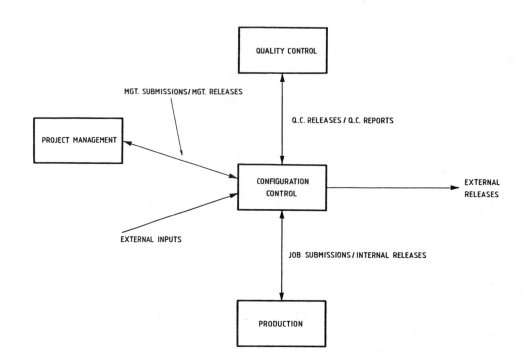

FIGURE 1 BASIC INFORMATION FLOWS

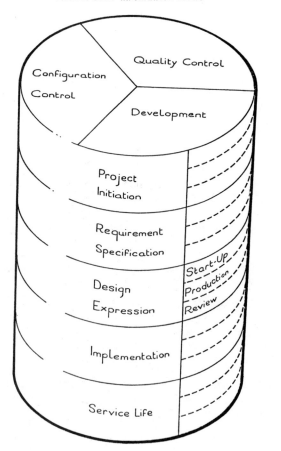

FIGURE 2 LIFE-CYCLE PHASES

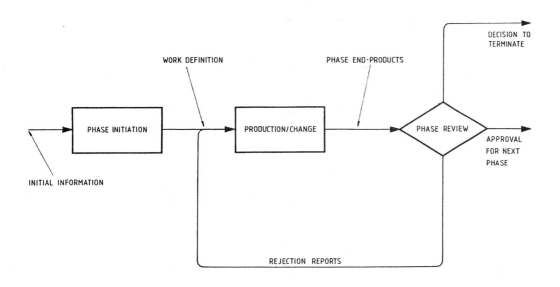

FIGURE 3 PHASE ACTIVITIES

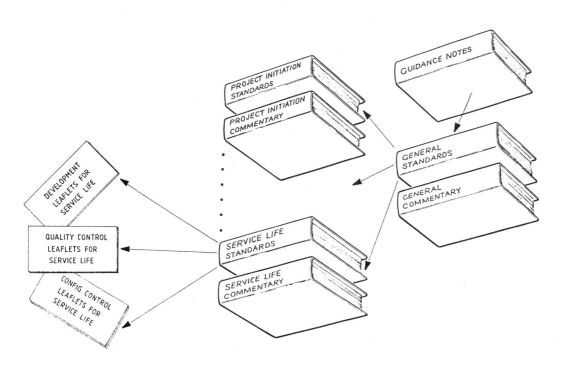

FIGURE 4 STANDARDS ORGANISATION

DISCUSSION

Maki: Often the pressure for better documentation seems to lead to none documentation. What approaches are you taking, in the preparation standards, to assist in the reduction of the bulk of documentation required?

Mullery: We believe that unnecessary bulk of documentation arises from the undisciplined presentation methods. We try to avoid this by ensuring that the definition of any item occurs in only one place. Other mention of the item name must be just a reference, rather than a definition.
However, there must inevitably be very much more documentation than is sometimes produced. Management should positively support the production of necessary documentation.

Lauber: I have the impression, that most of the problems, mentioned by the authors, of documentation and project management, can't be solved without automatic means. Do you agree with that?

Trauboth: I think we should aim at that, although i think we should start out disciplinning ourselves also in manual documentation and procedures first. Not proved methods should not be automated. Automated documentation and project management is also a question of cost. Attemts have been made to use existing management systems, which have been used in other area. The computer people have not used the computer, their own tool as much as they ought to.

Mullery: I want to emphasise that purchase, or production, of automated aids for a specific project, using the budget for that project is a false policy.
Few projects, viewed in isolation, could justify the expense of acquisition. Often, they could not justify use either. For example, if the cost of storing the software is large, or small project would find it uneconomic.

It is possible, however, to use the aid on several projects - either a sequence of projects or several projects in parallel. The cost-effectiveness then improves, because the cost of purchase and use is stared between projects.
Some simple projects may not need use of automated aids - simple production checklists may suffice.
If the end-products must later be used in a more restricted system, quality control may be introduced. Thus, the project may gradually build up quality, or show that it is worth throwing the end-products among and starting again, with better standards and use of automated aids.

FUNCTIONAL REDUNDANCY TO ACHIEVE HIGH RELIABILITY

M. H. Gilbert and W. J. Quirk

Atomic Energy Research Establishment, Harwell, Oxon, England

Abstract. In conventional protection and control systems, hardware is dedicated to perform a given function, and to protect against failure, this hardware has to be replicated. There is a trend to continue this practice into the field of computer-based protection systems. However in such systems, a single piece of hardware is responsible for many different functions, and redundancy at this level may not be the most effective way of achieving the desired high integrity of the resultant system. Furthermore, such an approach ignores the capability of computer-based systems to accomplish a large amount of self checking.

The important aspect of total system integrity is not the hardware but the functional integrity of the processes supported by the hardware. A fault in a processor can be looked upon as causing a change in the transfer functions of the processes it is supporting. By starting from a proper functional specification, it is possible to produce a system whose overall transfer function remains unchanged or at least acceptable in the face of a limited number of changed functional units. Thus the system integrity is maintained even if some of the processes do not implement the function specified - ie. contain design faults.

We explore a suitable specification technique in this paper, showing how it is possible to design such a system, and also how the individual functional units can be mapped onto the available processors in a decentralised way. A suitable architecture for such a system is discussed, based on a broadly conceived capability structure. Within this architecture, no process is statically assigned to a given processor element and the overall system behaviour is unaffected by the failure of a specified number of processors.

Keywords. Computer Architecture, Multiprocessing Systems, Reliability Theory, Special Purpose Computers, System Failure and Recovery, System Integrity, Fuctional Redundancy.

INTRODUCTION

The problem to be addressed is simple to state and understand; it concerns overkill, how to recognise it and how to prevent it. In a conventional hardware-based safety or control system, there is a more-or-less 1-1 correspondence between 'recognisable hardware boxes' or processing units and 'functions to be performed' or functional units. On the assumption that the processing units are initially functionally correct, the only thing which causes a system malfunction is a hardware failure of some sort. Replicating the processing units and voting between the results of ab-initio identical and correct units then increases the system reliability in that the overall system performance is unaffected by small numbers of failures in replicated sets of units.

At the risk of stating or restating the obvious, this approach to increasing system reliability does not immediately relate to computer-based systems. And this for two reasons: first, there is usually a many-to-one correspondence between functional units and processing units; second, it is much less obvious that any functional unit is correct ab-initio. One might add that there may well be subtle interactions between hardware and software which reveal themselves not immediately as a hardware error, but rather as a failure of some subset of the functional units mapped on to that processing unit.

However, this does not mean that computer-based systems are inevitably less reliable than hardware-based systems, for computers do have advantages in other respects. In particular, they offer a much greater checking-ability (either self-checking or one checking another) and because of the many-to-one correspondence between functions and processes, they allow a significant decrease in the amount of hardware needed to implement a given system, and a corresponding increase in basic system availability.

These advantages will be lost however if the computer is treated as a standard hardware device. Not only does straight redundancy not necessarily increase system reliability, but its introduction may cause the resultant overkill to prejudice adversely the overall performance achievable by the system. Overkill implies the sub-optimal allocation of system resources and therefore endangers the system availability, while the replication of similarly 'out-of-spec' units does not produce an 'in-spec' component.

RELIABILITY, RESOURCES AND FUNCTIONAL UNITS

While not wishing to get bogged down in detailed definitions and nuances of language, we need to be clear about what we are trying to achieve, and what is at hand to help us meet the objective. A system comprises a set of functional units which interact to produce the total system function. These units should be realised at specification or early design time. Resources are available to implement these functions at run-time. They will be realised during the system design. Both these sets will usually be redundant, in that not all the functional units need to be exercised or to even meet the specification for the system to behave in an acceptable manner. Further, some resources may become unavailable as time proceeds, and again, within limits, this should not adversely affect the system as a whole. Thus as a working but ad hoc definition let us define a system to be n-reliable if its <u>overall behaviour</u> is unaffected by the behaviour of any m of its resources or functional units, where m < n.

Now, although the visible parts of a system are the resources, the parts that really matter are the functional units. And for any particular invocation of a functional unit, it is the combination of functional unit and resource which will 'do' something.

In a standard triple modular redundant system, the use of three identical blocks and three voters is seen as a way of ensuring the correct majority result in the case of any single hardware failure. That is, the system is 1-reliable. We would suggest a slightly different slant to this; that the overall function of the 6 blocks is 'unaffected' by whatever function is implemented in any one of the blocks. 'Unaffected' is in quotation marks because clearly there is some effect, but in this particular case, it is the majority of the three outputs which is of concern, and the majority is truly unaffected by any one function.

So we see we must modify slightly our definition of n-reliable and say that a system S is n-reliable with respect to property P if P(S')=P(S) where S' is any derivation of S with less than n+1 of its active functional units changed. A simple example is in order, if only to clarify the above definition. Figure 1 shows in block form a system to raise an output alarm if any 3 or more of the 4 input lines are activated. Since any block could fail, we could have one output line raised spuriously or one output line not raised in an alarm situation.

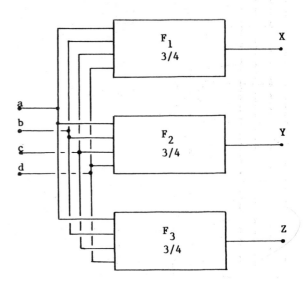

Fig. 1. A Triply-Redundant 3/4 Voter

Thus if P is the property

'<3 active inputs imply <2 outputs
 >2   "   "   "   >1   " '

then the system S is 1-reliable with respect to P.

Before we leave this example, let us expand it slightly to show another property of the functional unit approach.

If we consider the horrifying formulae:

$X = ((a \wedge b) \wedge (c \vee d)) \vee ((a \vee b) \wedge (c \wedge d))$
$Y = ((a \wedge c) \wedge (b \vee d)) \vee ((a \vee c) \wedge (b \wedge d))$
$Z = ((a \wedge d) \wedge (b \vee c)) \vee ((a \vee d) \wedge (b \wedge c))$

a realisation of which is shown in figure 2, then after a fair amount of analysis, we can discover that it also represents a 1-reliable system satisfying the functional requirements of figure 1. The interesting point is that there is no replication of any function in this implementation, although it does depend quite heavily on the symmetry of the situation. And because there is no replication, a software implementation would presumably be less prone to common-mode failures.

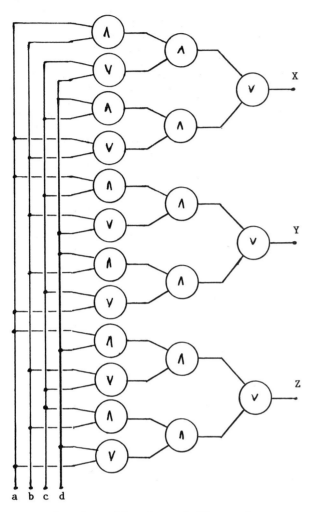

Fig. 2. A Realisation of Figure 1

In fact it is useful to stretch this definition very slightly. On one hand if a processing unit is recognised as faulty and the system takes some action in recognition of this fact, then after this 'reconfiguration' the system may again be tolerent of a single failure. Thus a system may well be p-sequentially 1-reliable meaning that up to p single errors may occur and the system still maintain the desired property. On the other hand, if a functional unit fails, it may fail repeatedly on each invocation. This is worse than just failing on a single invocation, and so we have a stronger requirement of reliability in time. We shall say that a system is 1*-reliable if the desired system property is maintained in the presence of a continually failing functional unit.

THE ALLOCATION OF FUNCTIONAL UNITS

The target we are trying to achieve is thus to produce a system which is both p-sequentially n-reliable and n*-reliable with respect to some safety property P. If this can be achieved at all, we need to know how to analyse 'functional' systems for failures, how to produce functional units and how to allocate them to processing units; the latter being the real hardware of the system. Again there are well established traditional approaches and perhaps Fault Trees are a good example. Can these be applied to computerised systems?

The usual answer is 'no', and this is because the computer and its software offer too many different failure states. One cannot even classify these states into a tenable number of equivalent states. What is wrong however is not the theory, but the magnitude of the analysis. However, this is almost entirely due to the requirements of general-purpose machines. The capabilities for a particular piece of software to be able to perform complex actions has led to the design of hardware in which any piece of software has such capabilities. This last statement is a little unfair - most machines around now have at least two levels of privilege, and some have more. But clearly no single-processor system can be 1-reliable, and most micro-processors - the obvious candidates for multi-processor machines - do not have such protection features. Indeed most mini-computers have the minimum possible in hardware and then rely on software at a particular privilege level to judge whether or not a request to that level from a lower privilege level is acceptable. In contrast, many micro-based systems do not have a vestige of such protection software.

Thus we see in general that the more complicated a machine - the more complex its architecture - the more failure states there are. What is needed is an architecture which reduces the number of such states. If a system can be made 1-reliable then since the single failure has no effect on the system, the exact nature of the failure is unimportant. In other words, the fault tree for the failure of functional units is just two-way branching.

Unfortunately this has only hidden the problem, because in general, a single hardware fault can fail many functional units. However, one thing which the microprocessor does allow one to do economically - is to allocate a single functional unit at a time to a resource. 'At a time' is very important, for we are certainly not talking about a static design in which anything in the older style hardwired systems is replaced by a less reliable modern counterpart. The significant point is that if there is only one function per resource at any one time, then one need not differentiate between resource failure and functional unit failure, and the fault tree analysis is now tenable.

Thus we are led to the position of requiring that functional units be allocated dynamically to processing units. Furthermore this must be done in a decentralised way or else 1-reliability cannot be achieved. Can such a system be produced?

FUNDAMENTAL 1-RELIABLE SYSTEM ARCHITECTURE

Let us recall the 3 fundamental building blocks identified in the earlier work on system specification (Quirk, 1978). These blocks were essentially <u>Clocks</u>, <u>Functional Units</u> and <u>Channels</u>. The semantic meaning of the SPECK model necessitated the allocation of the functional units to processing units; so this concept matches very neatly onto the 1-reliable machine concept. Clocks and Channels are in reality only special types of processes, and thus these too should match quite easily. However to achieve this, some extension of the normal functional definitions is required. Because to achieve 1-reliability, one must localise the effects of any error situation and this in turn requires that any action associated with error recovery is also local. In a system which is designed to be error-tolerent and perhaps also reconfigurable, reconfiguration and error handling should be seen as a normal part of the system operation. Thus the input and output spaces of each function should be extended to include a value 'failed', and the function definition expanded to be suitably defined on these values. Furthermore, the act of 'doing nothing' ought always to be an acceptable action for a single process, and should result in the corresponding output for that process being the value 'failed'. This can be achieved by having the associated output channel initialise itself to this value at process invocation. There are other ways of coding the outputs of functional units so that failures can be recognised. Another example, often used in conventional equipment, is to use an alternating output. Thus the boolean state <u>true</u> is represented not by a constant value but by a sequence such as 0,1,0,1,...

FAILURE DETECTION AND ASSOCIATED ACTION

With this very loose view of system components in mind, consider how failures could be dealt with in a 1-reliable system. It is dangerous to tread the dividing line between software failure and hardware failure, but we need some working rules for deciding how such a system should cope with errors. Self-checking of machines is quite feasible up to a point, but the problem then arises that any associated action to be taken on the self-discovery of a fault has to be taken by the failed machine. Thus there is no guarantee of this action being successfully taken. For a 1*-reliable system this implies that any self-checking mechanism must have associated actions with only local effects. In fact any sort of checking with associated global actions leads to problems (as we shall see) when the process with such global capabilities itself faults. We shall deal with this further below.

Of immediate concern is the following question: given that a fault has been detected, is it more likely to imply that the processing unit has failed, or that the functional unit has failed? If the former, the processing unit should be disabled, if the latter then it should not.

It is not clear how best to make this assessment, but the 'synchronised pair' of identical microprocessors (Cooper, Quirk and Clark, 1979) does satisfy the 1-reliable criterion; since only 1 failure at a time is to be catered for, then any discrepancy between the two processors (possibly after some retries) should lead to the pair being disabled. Since our processing units are rather more than just a bare processor, some similar hardware replication may be needed in the local memory, clock etc. The point to be made is that errors which are likely to be <u>hard</u> are the ones to detect at this level.

Other software detected errors could put the offending processor into a loop or halt, for example. this would then rely on the error being detected when the process over-ran its run-time bounds. If the channels are initialised to the 'failed' value for the process, then this value could be left there, as required. Thus the simplest possible action taken by the faulty process is acceptable to the system.

Fault annunciation is usually demanded of such systems, but we are wary of the value of such systems for several reasons. In the first place, the annunciation mechanism is an added complication to the system. Secondly, engineers invariably want to run special diagnostics to trouble-shoot systems. Thirdly, a time of failure is a time of stress for the system in that while attempting to recover or mask the failure, it is less tolerant of other failures. Thus it is not the time to be worried about annunciation. This is not of central importance to this paper, but we feel that equipment with regular service periods is probably better maintained by specialised diagnostic procedures rather than trying to make sense of a dated log of possibly spurious error reports.

THE ANALYSIS OF SINGLE SYSTEM FAILURES

The minimum necessary organisation of such a system must include the scheduling and allocation of functional units, and the data transmission between these units. Our 'processor pool' machine must therefore have as a primitive operation, 'allocate free processor' (say to the code in a particular piece of memory). To invoke a functional unit, one must allocate a free processor to that function. The obvious candidate for the allocator is that it is itself a function allocated to a processor, and this function has the capability to allocate free processor. One can now see how to decentralise scheduling, rather than have a continuously-running scheduler, we have a process which

- i) waits for whatever time delay is required
- ii) allocates the functions it controls
- iii) allocates a copy of itself
- iv) terminates, returning its own processor to the free pool.

Unfortunately, even this is not 1-reliable because if a scheduler dies, it cannot re-allocate itself. Thus, we need a watch-dog of some kind - again decentralised. The process as well as allocating the functions and a copy of itself, allocates a monitor. The monitor oversees all the processes invoked by this incarnation of the scheduler, excluding itself but including the next incarnation of the scheduler. This still has a window because the scheduler might fail before allocating the monitor. Monitoring monitors is starting to climb the dangerous and infinite pyramid of layer upon layer of checking processes. But a moments reflection on how we have decentralised the schedulers gives a clue. Just as the scheduler reschedules another copy of itself, so let the monitor look after another copy of itself. To be precise, we allow the monitor to oversee both the current and next invocation of the scheduler and its associated processes. Each monitor monitors the next incarnation of itself, and now there is no window.

Recall that our system is supposed to be 1-reliable; let us see if this design is so. There are four single failure modes for a particular allocation.

- i) an allocated function fails
- ii) the next scheduler fails
- iii) the monitor fails
- iv) a channel fails

Failure of a Scheduler. A scheduler fail clearly requires action by the monitor to reallocate a scheduler. Thus the monitor needs the capability to reallocate a scheduler but not a function. Without this capability, it should be impossible for a failed monitor to run amok among the other allocated processors, but one possible failure mode is for it to try and reallocate a scheduler when it need not. This would, of course, run two or more copies of the same scheduler. This is obviated by making its capability an 'abort-and-reallocate', for notice that aborting a scheduler does not harm the system provided that a new copy is almost immediately started. Leaving function reallocation to the next scheduler invocation is a good example of where the simplest possible form of error-associated action is utilised.

Failure of a Functional Unit. Functional units can fail in two ways: either they fail to produce an answer in the allotted time or they produce the wrong answer. In the latter case, it may or may not be recognisably wrong. If our system is 1-reliable, then it will not strictly speaking matter, but a system which can recognise errors can usually be made more effective. Failure to produce an answer can be spotted by the monitor for that function. If we assume that the communication mechanism can be informed of this failure, then all the monitor needs to do is to be able to abort-function. This unfortunately has severe implications for monitor failure.

Failure of a Channel. Inter-process communications are somewhat harder to deal with. Firstly, the input to the system and output from it will be down physical wires which are fixed. These channels cannot thus be moved around within the system. Even those which could be moved present somewhat of a problem in that copying data during a move may itself cause errors. The failure of a channel may be more catastrophic to the system than a failure of the function feeding it, because the memory of a channel makes its failure equivalent to a succession of failures of the function. This will be of no effect if the system as a whole is 1*-reliable.

Some latitude is available to cope with failed channels. Leaving aside those which are immobile, the others could be moved occasionally (a scheduled process would move them). But more interestingly, the monitors could check that reasonable data was being delivered. For example sequence numbers could be attached to data when written, and the monitor could then check that the correct number of different data values was provided at each channel read. A detected fault could then lead to the channel being aborted and a new one created. No data need be copied from failed channel to new one because of the 1*-reliability of the design. In this situation, rollback is not necessary so another potentially difficult area of system error recovery is bypassed. Indeed, one of the advantages of the channel-based architecture is the lack of replicated data which has to be kept 'in step' at all updates.

The immobile channels can of course only be dealt with by ensuring that sufficient redundancy exists in the presentation of data to the system so that single failures here do not adversely affect its operation. Notice here especially the value of the reading processes being able to determine the failure of the input channels to enable the function algorithms to make best use of the available data.

Failure of a Monitor. But what if a monitor fails? Since the monitor is not carrying out a crucial system function then in principle its failure has no impact on system performance. However since the monitor has some global capabilities, one has to worry about the misuse of these capabilities. The proposal here is that the combination of software and hardware techniques reduce the probability of such misuse to the 'inconceivable' level. The capabilities granted to any process in a system should be just the minimum required by that process, so even a monitor is not free to do anything and everything. The software will be well trusted - probably proved - and by suitably coding the capabilities, the probability of a corrupt capability being accepted as another capability should be made very low. Similarly the capability mechanism will be a well trusted hardware construction. Finally, the monitor should have the capability to abort the single monitor it itself is monitoring. Thus any attempt to 'run amok' by one monitor will be arrested by its overseer.

CONCLUSION

Let us stand back at this point, for a paper of this nature should not be a detailed design document. Clearly there will be problems producing an architecture which meets some of this requirement just mentioned, but these are largely conventional hardware problems. What we hope to have indicated is that with a broad concept of capability mechanisms, one can at least envisage architectures which allow provable general statements about the system behaviour to be made. The possibility of achieving a 1*-reliable design seems to be real enough providing that the basic system specification allows one to recognise the consequences of a different function being implemented in one of the functional blocks. TMR is a standard example of this. The results of such a specification are that the inherent redundant parts of the system are not needlessly replicated, but that the redundancy neccessary to achieve a given level of fault tolerence can be implemented. A look at a specification will often reveal the typical pyramid shape of much redundancy at the input tailing off to very little at the output. With a SPECK-type specification, the consequences of a function change at any point can be traced, and a hand analysis or simulation can establish whether or not there is 1- or 1*-reliability with respect to this function.

We are currently working on the problems of producing a suitable architecture, and we can say little more until this work is more advanced. We suspect it will be a case when a small amount of spectacular testing - for example pulling boards out of a working machine without disturbing it - will seem more convincing than pages of detailed argument. However, surely the important point to bear in mind is that the ability to be able to offer some sort of proof of system integrity is crucial to licensing authorities, and that such an unconventional system as we have been speculating about here is only justified if it offers some advantage over a more normal approach. Problems such as the importance of common-mode failures are all too real in systems such as these, but these difficulties do not seem to be any more acute in such systems than in many others. Finally, this does seem to be an approach in which the full and powerful potential of multi-processors can be achieved in a controlled and flexible manner.

REFERENCES

Quirk, W. J. (1978). The automatic analysis of formal real-time system specifications. AERE-R 9046.

Cooper, M. J., W. J. Quirk, and D. W. Clark (1979). Specification, design and implementation of computer-based reactor safety systems. AERE-R 9362.

DISCUSSION

Walze: You described a system using redundancy in a dynamical way, i.e. dynamic allocation of hardware modules to functions. Other approaches are cold-redundancy or stand-by-redundancy. My impression is that your approach needs much more organization efforts.

Gilbert: The redundancy we described is on the functional level, therefore there is the possibility of optimization and failure recovery.

Lauber: When you talk about optimization criteria you mean hardware cost?

Gilbert: Not necessarily, we expect to achieve optimizaion in total, that means hardware and software.

COMMUNICATION PROTOCOLS FOR THE PDV BUS IN NETWORK REPRESENTATION

T. Grams and M. Schäfer

Brown Boveri and Cie, Mannheim, Federal Republic of Germany

Abstract. The PDV bus is a bit-serial data transmission system for process control systems with distributed stations. The connection between the stations is established by a continuous transmission line or bus. Data exchange takes place by time-division multiplexing (half-duplex operation).

Proper handling of the exchange of messages via the bus is taken care of by the communications protocol. The function of the protocol is to adapt transmission optimally to the time requirements, i.e. to ensure a high efficiency. In addition, the protocol must afford high transmission reliability.

The communications protocol must tolerate faults in that it prevents dangerous system conditions such as conflicts, blocking and loss of messages under all circumstances upon the occurrence of the faults to be tolerated. For verifying the fault tolerance, a properly formatted and unique description of the communications protocol is necessary. The basis of such a description is provided by the Petri networks which are particularly suitable for the representation of "communicating processes". A suitable graphic representation of these networks is introduced. The most important procedures of the PDV bus communications protocol are formulated in this network representation. This provides the basis for a detailed analysis and for the verification of the fault tolerance.

Keywords. Communications control applications, distributed systems, industrial control, PETRI-nets, process control.

INTRODUCTION

For the process control of complex processes with distributed stations the conventional centralized system structure leads to an extremely high expenditure for jumpering operations as well as high costs for cable and cable laying.

The technology of the large-scale integration with its large-volume memories, microcomputers and special application-oriented circuits offers the possibility of a nearly unlimited distribution and extension of capabilities. Higher reliability, projectable availability, improved user-oriented configuration, better utilization of the communication lines, as well as the upkeeping of extendibility even over longer periods can be realized by this technology. The demand for an as linear as possible dependance of the costs on the degree of extension leads to the introduction of terminal line and bus systems with regard to data transmission.

Furthermore, if one regards the increasing specialization of the manufacturers in certain control components it becomes more and more difficult for the user to buy control systems for large plants on a turnkey basis from one firm. The components of different manufacturers can hardly be linked.

The standardization of a process bus system with a wide application range can provide a remedy for this problem. Serial bus systems with standardized interfaces are of special importance because they offer the possibility to realize low-cost transmission facilities.

In the following the serial bus system worked out within the framework of the PDV project[1] is taken as a basis. This system represents with its effective bus functions (for instance writing, reading, cross communication, demand handling) a technically and economically purposeful means which contributes to the construction of many data transmission systems (1).

---

[1] The project "Process Control with Data Processors (PDV)" is promoted within the framework of the data processing programmes of the German Government and coordinated and attended by the Nuclear Research Centre, Karlsruhe GmbH.

It is the aim of this paper to furnish a properly formatted description of the communication protocol on the basis of the Petri networks and to verify important qualities of the protocol. The paper contains a brief presentation of the PDV bus in accordance with (1).

CHARACTERISTICS OF THE PDV BUS

The PDV bus is a bit-serial data transmission system for process control systems with distributed stations. The connection between the stations is established by a continuous transmission line or bus. Data exchange takes place by time-division multiplexing (half-duplex operation). Each station of the system consists of a bus coupler and a data terminal. The data terminals are connected to the bus through a standard interface, the Serial Digital Interface (SDS). This interface is independent of the line design (for instance pulse transmission, frequency modulation, light transmission). Bus coupler and line must be physically matched and form the transmission medium (Fig. 1).

The basic functions of the SDS are

1. Representation of binary digits
2. Designation of beginning and end of a bit sequence by frame signal
3. Indication of detected transmission errors by a signal "data signal quality" to the addressee.

In simple systems a master station transmits command messages and expects a reply message from the station addressed in the command message. The stations monitor all command messages and transmit, if they were addressed themselves, a reply message.

A cross communication is provided for the message exchange between two stations. It enables a station – after having been authorized by the master station – to directly exchange messages with another station. This station is then temporarily assigned the function of transmitting command messages.

Message Format

The communication element of a message is an 8-bit field (byte). Depending on its position within the message this field can be an address field (A-byte), a functional field (F-byte), a CRC (cyclic redundancy check) field (S-byte) or a data field (D-byte). In all cases 2 bytes are secured by one CRC byte (Fig. 2). The address field A designates the station address, in the command message the destination address and in the reply message the transmitter address.

The functional fields of command and reply message transmit different contents. They contain information for transmission and process control and especially facilitate the identification of the message type (command or reply message). The exact coding is described in (1).

Message Exchange

Messages are always exchanged in cycles between stations. A message cycle consists of command and reply message. The data can be transmitted in the command message (writing) as well as in the reply message (reading) (Fig. 3). Basically each station must monitor all messages in order to be able to know when it is addressed. Figure 4 shows a message cycle during a message exchange between two stations (cross communication, in short: C.C.).

Error Detection and Recovery

Error detection measures are carried out on different levels of data transmission (error detectors for the line signal, evaluation of the CRC field, reasonableness checks).

Error recovery is in all cases carried out according to the principle: "If in doubt do nothing". This means: an incorrectly received command message is not acknowledged nor executed. The calling station then repeats its command message. If the reply message is faulty the command message is also repeated and thus the reply message is transmitted again.

The consequence of this principle is that transmitted messages must be stored in the transmitting stations until it is known for sure that a repetition is not needed anymore. This is the case when the station has received and answered error-free a new command message directed to it.

Figure 5 shows the Watch Dog mechanism required for message exchange and cross communication. The individual time periods are determined and monitored by the master station or the calling station.

THE FUNCTIONS OF A COMMUNICATION PROTOCOL

If the interfaces have several data terminals conflicts (simultaneous transmission of several data terminals) and inadvertent blocking of data terminals must be avoided. Another function of the communication protocol is the avoidance of message losses and message duplication.

Furthermore the protocol is to allow an optimum adaptation of transmission to the time requirements, i.e. to ensure a high efficiency.

The main consideration here is the safety of the communication protocol, i.e. the capability to avoid dangerous consequences (message losses, conflict) of certain unavoidable errors (2).

The following errors occur at the SDS and must be tolerated:

1. detected transmission errors
2. omission of messages (receive frame does not respond although message is being transmitted).

Before the error tolerance of the PDV bus communication protocol can be verified an properly formatted description of the protocol must be made. The Petri networks serve as a means of description.

DEFINITION OF THE NETWORK REPRESENTATION

The Petri networks are a generalization of the state diagrams. Basically state diagrams are used for the description of systems which carry out algorithms. Each node of the diagram corresponds to a certain process state and thus to a certain point of the algorithmic procedure. The algorithmic procedure corresponds to a mark passing along the diagram. Passing into the next state is triggered by events whose occurrence may depend on conditions.

The state diagrams are well suited for describing processes as they take place in the individual data terminals of a bus system. However, the interrelations between the different processes, i.e. the synchronization of the processes by interface messages, cannot be seen from the graphic representation (4).

The Petri networks go beyond that by facilitating the joint representation of several state diagrams and the synchronization conditions between them in a uniform and compact network representation. According to this several state nodes can be simultaneously marked in such a network, contrary to the state diagram.

Figure 6 shows the essential elements of the Petri networks. The connection of the state nodes is carried out via events. An event has any number of nodes as inputs or outputs. The alteration of state marks is done according to the following rules: the markings of all inputs are a necessary (however, not a sufficient) requirement for the beginning of an event. The occurence of an event results in the fact that the marks go from the input side to the output side. The total number of the marks will change if the number of the input nodes and output nodes are not identical.

A somewhat more modified Petri diagram is used for the network representation of the communication protocol, i.e. of the different transmission procedures. In order to arrange the representation more clearly the network element "conditional transition" is introduced (Fig. 7).

Effect of the "conditional transition": A is a precondition of transition B. With the input node marked, event A is always possible. Transition B can only take place simultaneously with transition A. A and B are transitions of different state diagrams. This furnishes an easily remembered representation of the effect of interface messages between transmitter and receiver taking into account the permissible errors.

As an additional simplification the rectangles (events) to or from which only one arrow points can be despensed with (Fig. 8). According to this agreement state diagrams really are special Petri networks.

NETWORK REPRESENTATION OF THE COMMUNICATION PROTOCOL

The following agreement is made as far as the network representation of the communication protocol is concerned: interface messages are written in capital letters. Device messages are written in small letters.

A "conditional transition" (see chapter 4) as transition condition means that a message must be transmitted and decoded error-free and that the received address corresponds to the expected one.

Message Exchange

In a centrally oriented control system the simplest form of message exchange consists of command messages of a master station and of reply messages of the individual data terminals. This is also the basic procedure of the Serial Digital Interfaces. With both messages data can be transmitted simultaneously (writing, reading). Figure 9 shows the network representation of this procedure and Fig. 10 a survey of the states. At the beginning only the states EMP are marked (starting condition). If a command message is to be transmitted the master station moves from the passive state EMP into the transmitting state AUS.

After the transmitting procedure the station waits in state AWT until a reply message of the addressed data terminal is received and then goes back into the passive state. If no error-free reply message is received in the time interval TAN there will be a transition to the error state STG. Subsequently a repetitive command message is transmitted in WAS. This cycle is repeated until a reply message is received.

If a data terminal receives a command message addressed to it it will move from the passive state EMP into the state NAV where a reply message is prepared which is subsequently transmitted in ANS. The reply message is still stored in the passive state after the termination of the transmitting procedure and is repeatedly sent on receipt of a repetitive command. After the activation or normalizing of the reply function a pseudo message (for instance status message) is ready as a reply, and not the message to be transmitted, since this pseudo message is replaced on the transmission of the first command message.

Cross Communication

The cross communication is especially characterized by the fact that the function "command" is not reserved to only one station but several stations are equipped with it. Additional basic functions can also be provided for in the stations. Only the basic function "assignment of the command function" is allocated to merely one station, the master station. The latter determines which station the actual command station is in the system. It makes use of the cross communication procedure and assigns to one station (with command ability) the command function either cyclically or on request. This station is then able to exchange data with any other station in the system. After this operation is ended the station returns the command function to the master station.

Figure 11 shows the network representation of the data exchange during cross communication. The assignment function for it shows Fig. 12. The states are shown in Fig. 13, as far as not already shown in Fig. 10.

If the command function is to be assigned to a station the master station moves into state ZUS. After the transmitting procedure is ended it waits in state WTZ until the commanding data terminal returns the command function with the message "END C.C.". If this does not happen within time TQV there is a disturbance.

The network representation of the command function is slightly modified. The transmitting of a command (AUS, WAS) is only carried out on being addressed with the message "START C.C.". After the message exchange has been ended (see Fig. 11) a transition to EQS takes place where the global command "END C.C." is transmitted. Not until then the data terminal proceeds to the passive state EMP.

If no command message is to be transmitted when an assignment is received the global command "END C.C." will be transmitted immediately (transition from EMP to EQS).

It is a precondition for the functioning of the cross communication that always only one command function is active in the whole system.

ANALYSIS OF THE PROTOCOL
(VERIFICATION OF THE FAULT TOLERANCE)

In order to investigate the protocol it is practical to integrate the Petri networks of the individual procedures in a compound state diagram. Compound states correspond to the simultaneous occurrence of states in different state diagrams. The compound state (Z1, Z2) = (EMP, EMP) consists of the states EMP in data terminal T1 or T2. The transitions between the compound states are derived from the Petri network step by step. For this process the compound state (EMP, EMP) is taken as a basis and all possible transitions are designed in succession.

In the following the analysis is to be made with the procedure "message exchange" as example. Figure 14 shows the compound state diagram of the procedure (for explanation of the state designations see Fig. 10).

a.) Blocking

As can be seen from the diagram the blocking of a function in a state is not possible. For from each attainable state a path leads back to (EMP, EMP). On the occurence of a permanent error an infinite loop might be possible, however, it can be detected by the command function.

b.) Conflicts

Conflicts may occur when the compound states (AUS, ANS) or (WAS, ANS) are assumed. However, this is only possible in the case of state (WAS, ANS) by a faulty time condition TAN (dotted transitions). However, this can be avoided by the right selection of TAN.

c.) Message losses

Message losses can be avoided by not replacing a message in a station by a new one until the old one has been properly received and acknowledged. And this corresponds to the request that state NAV of the reply function may only be attained via the compound state (EMP, EMP). This can be easily verified in the diagram since state (AWT, NAV) is only reached via (EMP, EMP). All the other compound states with NAV are never reached

The protocol does not have one of the abilities mentioned in chapter 3: it cannot avoid the message duplication. This can also be seen in the network representation. A message duplication during writing will occur if the command function does not take state AUS between two transitions from EMP to ANS in the reply function. Figure 14 shows that this is the case with loop (AWT, EMP), (STG, EMP), (WAS, EMP), (AWT, ANS), (AWT, EMP). A similar rule applies for reading: message duplication will occur if state NAV in the reply function is not taken between two transitions from AWT to EMP in the command function. This is realized in the loop (EMP, EMP), (AUS, EMP), (AWT, EMP), (STG, EMP), (WAS, EMP), (AWT, ANS), (EMP, EMP).

Message duplication can be avoided for example by message counters. These can be more easily realized on higher protocol level (user protocol) than on the level of the general communication protocol. In many cases such devices can be dispensed with because message duplication is not disturbing (5).

REFERENCES

(1) Buxmeyer, Deck, Hausmann, Janetzki, Möhl, Walze (Oktober 1978) Serielles Bussystem für industrielle Anwendungen unter Echtzeitbedingungen (PDV-Bus) Kernforschungszentrum Karlsruhe, PDV-Bericht, KfK-PDV 150

(2) Heckmann, H. (1977) Sicherheits-Definition und Bewertung insbesondere für Steuerungen INTERKAMA-Kongreß 1977, Springer-Verlag Berlin

(3) Wendt, S. (1974) Petri-Netze und asynchrone Schaltwerke, Elektronische Rechenanlagen 16, 6, pp. 208-216

(4) DIN IEC 66.22 (1976) Elektrische Meßtechnik Byteserielles bitparalleles Schnittstellensystem für programmierbare Meßgeräte. Entwurf

(5) Gray, J.P. (1972) Line Control Procedures Proc. IEEE 60, 11, 1301-1312

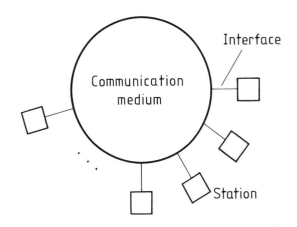

Fig. 1 Data transmission system

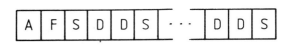

Fig. 2 Message format

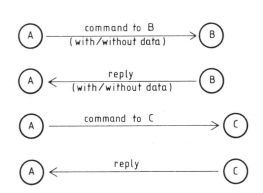

Fig. 3 Sequence schedule of the message cycle (A: active master station, B and C: stations)

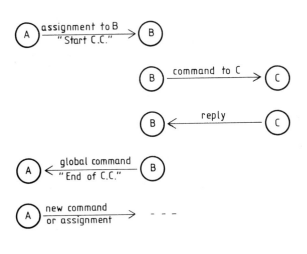

Fig. 4 Sequence schedule cross-communication (C.C.)
A: active master station,
B: commanding terminal,
C: replying terminal

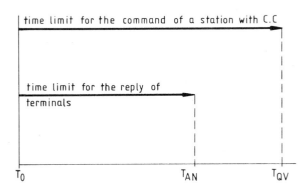

Fig. 5 Watch Dog mechanism

Net elements	Representation	Effect
State, State node	 ANS	system passes over to ANS and leaves this state on certain conditions
Event, transition condition	 E	event E only takes place, if all state nodes leading to E are marked

Fig. 6 Basic elements of PETRI-nets

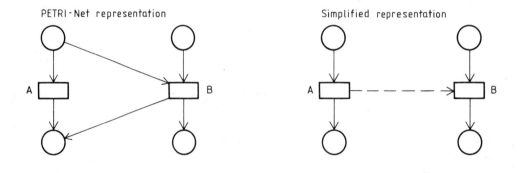

Fig. 7 The net element "Conditional transition"

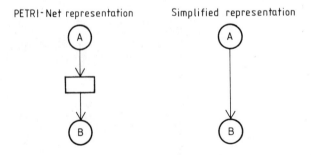

Fig. 8 PETRI-Net and state diagramm

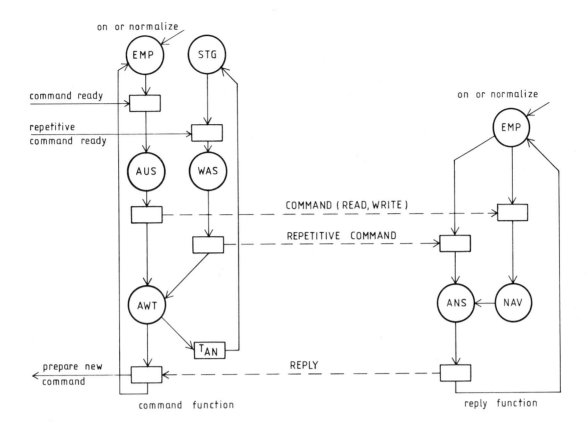

Fig. 9 Net representation "message exchange"

Abbreviation	Description
EMP	transmitter/receiver in passive state
AUS	command transmitting state
AWT	command function in wait-state
STG	command function error-state (reply faulty or reply not received)
WAS	repetitive command transmitting state
ANS	reply transmitting state
NAV	prepare new reply

Fig. 10 States of Net representation "Message exchange"

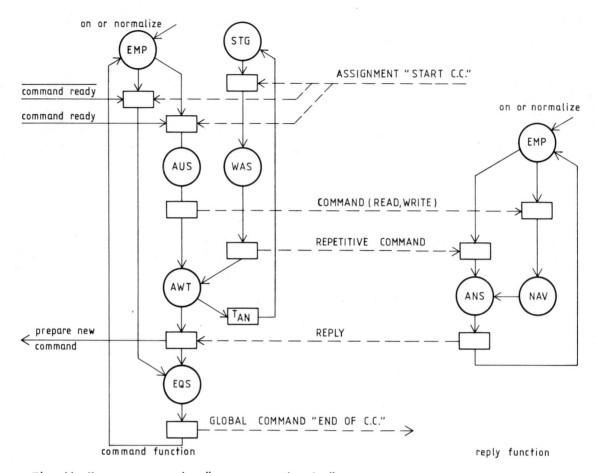

Fig. 11 Net representation "cross-communication"

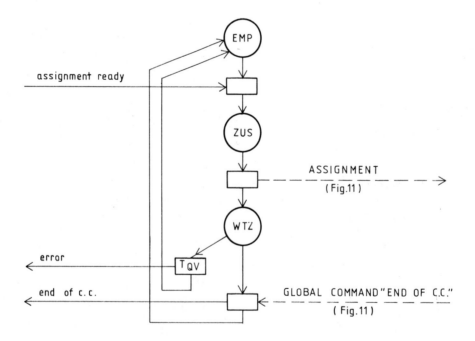

Fig. 12 Net representation "Assignment function"

Abbreviation	Description
ZUS	assignment-transmitting-state of the master station
WTZ	wait-state of the master station
EQS	commanding terminal returns command function (message "END OF C.C.")

Fig. 13 State description of the net representation "Cross-communication"

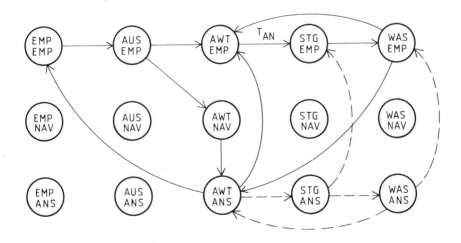

Fig. 14 Compound state diagram, procedure "exchange of messages"

DISCUSSION

<u>Walze:</u> Can you recommend Petri-nets for other communication systems?

<u>Grams:</u> Petri-nets are well suited for the graphical representation of small systems with low complexity in state diagrams and a high degree of communication and synchronization between the systems.

<u>Ehrenberger:</u> Do you know anything about the failure behaviour of your procedures and of the associated hardware? If you know, what is the error probability of transmissions carried out for certain data to be transmitted.

<u>Grams:</u> For this configuration here we have made some calculations but the results aren't available yet. Transmission errors can be detected for example by cyclic redundancy checks or plausibility checks.

<u>Ehrenberger:</u> Is the method related to a particular kind of cable or can it also be used e.g. with light conductors? Do you need some particular band-width and what is the transmission-speed?

<u>Grams:</u> The PDV-bus is independent of the type of transmission line. The SDS is introduced to interface with the various transmission media.

<u>Pau:</u> How do you detect interface failures and how do you account for these?

<u>Grams:</u> The protocol is designed to deal with transmission failures. Hardware failures that result in transmission failures are treated alike.

<u>Pau:</u> Isn't there too little space for status information in your protocol?

<u>Grams:</u> If you need more status information in your protocols you can locate it in the F-field and the D-fields of the message. The message can be of variable length.

<u>Parr:</u> It seems to me that introducing your protocol, a local failure can lock out the whole bus e.g. by permanent repetition of a request until a correct response is received.

<u>Grams:</u> This is a certain level of protocol. There are other layers of protocol where the user can handle these problems.

<u>Parr:</u> How can you accept an implementation with one single central master in a distributed system?

<u>Grams:</u> The master station has only very simple functions, all stations are able to enter the bus in the same way. Beyond that there are procedures for cross-communication and master transfer.

<u>Gilbert:</u> Is it true that if one command in your protocol is lost the same reply will be repeated again?

<u>Grams:</u> Yes, this is a case where message duplication may occur. You can avoid that by message counters etc. in your user protocol. It is not provided on this level of transmission protocol.

<u>Quirk:</u> Is it possible for the user to abort an outstanding request and thus unlock the repetition cyle in your diagram?

<u>Grams:</u> Yes, this can be done by a normalizing command in every state.

<u>Maki:</u> There are lots of protocols becoming standards in communication networks e.g. HDLC, why was it necessary to invent a new one?

<u>Grams:</u> The PDV bus protocol is a simple one, it needs no extra storage and counters and is very easy to implement. These may be some reasons for it to be preferable to higher level protocols as HDLC. The application fields of the protocols are different: The PDV bus is designed with respect to process control applications, fulfills high safety requirements and permits short reaction times whereas the HDLC is optimized with regard to the transmission of long message blocks.

<u>Additional statement by Walze:</u>
The PDV bus is not longer an approach it is now implemented by Hartmann & Braun (FRG) and it is a candidate system for national and international standardization.

SOFTWARE DIVERSITY IN REACTOR PROTECTION SYSTEMS: AN EXPERMENT

L. Gmeiner and U. Voges

Kernforschungszentrum Karlsruhe GmbH, Institut für Datenverarbeitung in der Technik, Postfach 3640, D-7500 Karlsruhe, Federal Republic of Germany

Abstract. Since simple duplication of software does not pay off for increased reliability, a special kind of redundancy has to be used: diversity. This paper illustrates different kinds of software diversity and describes an experiment in which software diversity was applied in a prototype implementation of a reactor protection system. In the course of this implementation the errors were reported and classified. The advantages of diverse programming, concerning the aspect of easier validation of the software, are explained.

Keywords. Nuclear reactor protection system; redundancy; software diversity; software error classification; software reliability; software validation.

INTRODUCTION

In computer applications with high reliability and availability requirements, like applications in the process-control area, hardware as well as software has to have the ability to detect existing errors and to react to error situations. This paper will deal especially with one ability to increase the reliability of software, with diversity. The first chapter reports on related work and gives some definitions. The next chapter explains the reactor protection system and the use of diversity within it. Following this, the experiment and its results are described. Finally possible achievements and experiences are summarized.

CURRENT SITUATION AND RELATED WORK

For hardware there exists already a broad knowledge of and much experience in the use of redundancy. Hot and cold (stand-by) redundancy can be distinguished. Cold redundancy has some equivalence in the software-area, the recovery-blocks (Anderson, 1977). Here the programmer has the opportunity to define several alternative program parts for the solution of a problem. If one of the alternatives delivers a result which does not pass an acceptance test, it can be assumed that an error occured during the computation. The next alternative is chosen or at least an error routine is activated.

By means of language terms a recovery block can be described by:

```
ensure      <acceptance test>
       by   <first alternative>
else   by   <second alternative>
            :
            :
else   by   <n-th alternative>
else        <error handling>
```

Normally, hardware redundancy is realized by simply duplicating identical components or even systems. The different components are operated in parallel and a majority voter decides on their results (m-out-of-n-systems with voter). Since the single duplicated components are developed according to the same specifications and design logic, only aging-errors and production-failures can be detected by this technique. If also logic design errors shall be detected, the design has to be duplicated, too. This kind of redundancy is then called diversity. E.g. in hardware, different components with identical functions, are used or different physical effects are measured for the calculation of a certain quantity.

In the KTA-rule 3501 (1977) the hardware diversity is defined as the use of redundant components of different architecture or different mode of action. Accordingly we want to give the following definition of software diversity:

Software diversity is the use of redundant program code, which is generated in a different manner starting from one specification.

The definition shows, that diversity is a special case of redundancy. Other terms for software diversity are 'n-version programming' (Avizienis and Chen, 1977) and 'distinct software' (Fischer, Firschein, and Drew, 1975).

REACTOR SAFETY SYSTEMS AND SOFTWARE DIVERSITY

Generally reactor safety systems are designed as hardware m-out-of-n-systems. The safety system dealt with in this paper (see Fig. 1)

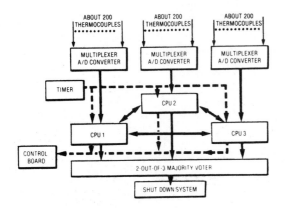

Fig. 1. Scheme of the computerized reactor protection system

was designed for supervising the fuel element temperatures (Jüngst, 1976). The three computers of this system shall be operated with the identical software, due to cost reasons. These costs divide up into software generation costs and validation and verification costs. With increasing availability and reliability demands especially the validation and verification costs increase. We think that the use of diverse programs can limit this increase. On first sight one might have the impression this is wrong, especially if one looks at the software generation costs, which are multiplied by the number of diverse programs. But this is at least equalized, if not even improved, through the decreased probability of common mode errors. During verification and validation activities, not as much attention has to be given to this kind of error as would be necessary in systems which have identical software in all computers. Therefore the verification costs are lower. Additionally the single programs need not to have a reliability as high as in single installations, since the necessary system reliability is achieved through the 2-out-of-3 system.

Different grades of diverse programming can be distinguished, which are listed here with increasing complexity:

- diverse implementation of an algorithm, using different programmers, but the same language;
- diverse implementation of an algorithm, using different programmers and different languages;
- diverse implementation of an algorithm, using different programmers, different languages and different computers.

The first grade of diverse programming was the basis for an investigation by Avizienis and Chen (1977). The second grade will be described here. The third grade of diversity does not only concern software, but also hardware. It would be interesting to have a project for this diversity.

In general it can be stated that the probability of common mode errors is decreasing if the level of diversity in the above list is increasing.

DESCRIPTION OF THE EXPERIMENT

Implementation

The starting point for the single implementations was a common specification. In order to have specifications, as precise, consistent and complete as possible, we have chosen a formal specification method, which is similar to the input-process-output approach (Boehm, 1974). The total program is described as a set of mutual linked processes. Each of these processes consists of input, output and states, and each process state is defined by the actual values of the basic set of variables. One aspect in the specifications is quite important. In addition to the synchronization points at the end of each program cycle a set of 'internal checkpoints' was specified, where some intermediate results of the separate implementations can be compared. These checkpoints gave the possibility - as will be explained later - to detect certain errors in the implementations which would not have been detected by simple comparison of the final program outputs.

For the implementations we used the different languages IFTRAN (1976), PASCAL (Jensen and Wirth, 1974), and PHI2 (1977). These languages were chosen since they were available on our SIEMENS 330 computer, and on the other hand these languages allowed to a fairly well degree to follow the programming guidelines (Voges and Ehrenberger, 1975) which were postulated for this project.

The used formal specifications proved to support the implementations in the languages IFTRAN, PASCAL, and PHI2. It was often possible to transform the formal specifications quite naturally into the code of the programming language.

IFTRAN is a preprocessor for structured FORTRAN, PASCAL is a well-known high-level language, and PHI2 is a macro assembly language which contains structure macros for 'if-then-else', 'case', 'do-loop' etc. Through this use of languages of different levels (machine level to high level) a special kind of diversity is realized, which decreases the probability of common mode errors furthermore.

The validation methodology (Geiger and others, 1979) which was the basis of this experiment, contained not only constructive methods, but also analytical methods, and here mainly an intensive program test. These tests were carried out in several phases. First the program was tested by the programmer himself. Afterwards the program was retested by another person, according to predefined testing criteria. Besides diverse programs an additional aspect of diversity was realized by this technique: a diversity of the staff during the program development cycle.

As far as possible, the tests made use of automated tools like the test systems RXVP (1975) and SADAT (Amschler and Gmeiner, 1977) and an automatic result comparisor.

A summary of some characteristic implementation quantities is given in fig. 2. The experiences gained in this experiment will be described in the next chapter.

	IFTRAN	PASCAL	PHI2
Code size (words)	21251	13463	14205
program lines without comments	1077	783	1606
runtime normalized	1,21	3,65	1

Fig. 2. Characteristic figures of the three implementations

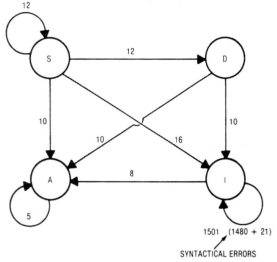

S = SPECIFICATION
D = DESIGN
I = IMPLEMENTATION, MODULE TEST, INTEGRATION TEST
A = ACCEPTANCE TEST (BY APPROVING AUTHORITY, BY USER)

Fig. 3. Correlation between error cause and error detection (arrow-head) concerning the phases of the program development cycle.

Experiences

Besides the aforementioned use of tools during testing another main effort was to write down a detailed error report and analysis for the total experiment. All errors which were detected during the project were documented on report forms and analysed at the end of the project (Gmeiner, 1978). In the following some of the main results of this analysis shall be explained.

Fig. 3 shows in form of a graph the relation between error cause and its detection (arrow-head) concerning the phases of the program

development cycle. This demonstrates that the specifications play a dominant role. Most of the errors are closely related to the specifications, and the errors induced during this phase become effective in all the following phases (mostly by ambiguity and misinterpretation of the specifications).

Fig. 4 gives in form of a matrix the relation between the error types and their detection methods. Out of the used methods especially the method of 'automatic result comparison' is interesting, since it is a characteristic feature for the usefulness of diverse programming. As already mentioned in the last chap-

error-type \ detection-method	automatic result comparison	manual result comparison	desk check	automatic test systems	run time system (PASCAL)	(macro-)assembler, linkage editor	machine based debugging tools (Tepos, ...)
incomplete specs	2	3	16	-	-	-	-
ambiguous specs	3	2	10	-	-	-	-
logical error	4	3	11	-	-	-	10
typing error	1	2	-	4	-	1	-
interface error	-	-	1	2	-	-	-
other implementation error	3	-	6	4	-	-	7
error in testframe	-	-	1	-	-	1	-
error in system software	2	4	-	-	3	2	1
other errors	3	13	-	-	2	4	1

Fig. 4. Relation between error type and error detection method

ter, the specifications defined some intermediate checkpoints, which were used during testing. An automated comparison program checked the three implementations in order to see whether the intermediate results at these checkpoints were identical. Most of the errors which were detected by this method were due to ambiguous specifications. The following example shall explain such an ambiguity.

Specification:
$$m^{min} \leq m_i, m_j \leq m^{max} \quad \forall\ i=1,\ldots,205,\ \forall\ j=1,\ldots,3$$

Interpretation (1):
$$m^{min} \leq m_i \quad \forall\ i=1,\ldots,205$$
$$m_j \leq m^{max} \quad \forall\ j=1,\ldots,3$$

Interpretation (2):
$$m^{min} \leq m_i \leq m^{max} \quad \forall\ i=1,\ldots,205$$
$$m^{min} \leq m_j \leq m^{max} \quad \forall\ j=1,\ldots,3\ .$$

A further advantage of diverse programming is the simplification of the test process. Normally for all test input data the corresponding test output data have to be computed more or less manually. If a diverse implementation exists, the results of the two implementations can be checked against each other, and if one neglects common mode errors, all errors resulting in different intermediate or final results can be detected. Thereby it is possible to check the diverse programs just by automatically generating large amounts of test data and by comparing the results whether they are identical or not. Only if discrepancies are detected, it is necessary to check which of the results is incorrect.

From the total of the detected errors 18 errors ($\triangleq$ 14 %) were detected by the automatic comparison program. This seems to be a fairly high number, especially if one takes into account that all the other testing methods were used before the three implementations were compared with each other. Therefore these errors were probably the hard-to-detect ones.

To conclude it can be said that the automatic comparison program was a valuable tool for the comparison of the large amount of data, which would have not been possible, at least as error-free, by hand.

CONCLUSION

The probability of a common mode failure (the same error identical in all implementations) is very low if the teams work independently, the languages have different levels (assembly language and problem oriented language, e.g.) and the compilers were not designed by the same people. The most probable reason for a common mode failure is an error in the specifications. Therefore much effort has to be put into correct, consistent and complete formal specifications. However, there is a problem in detailing the specifications too much: Detailed specifications lead to similar internal program structures of the different implementations. Thus the advantage of diverse programming gets partly lost.

According to our experience one of the main problems in diverse programming is the synchronisation of the different programs, especially if the programs have intermediate checkpoints. These checkpoints have to be designed explicitly, detailed and very carefully. They proved to be valuable especially since we detected a few errors, where final results agreed and intermediate results disagreed. In these cases we have detected some hidden errors, not activated directly by the test cases.

Furthermore, the experiment showed that diverse programming detects certain errors which are unlikely to be detected with other methods.

We think that this technique of diverse programming may lead to a considerable reduction of the test and validation effort necessary for safety related software like reactor protection systems.

REFERENCES

Amschler, A., and L. Gmeiner (1977). SADAT 2 - Ein System zur automatischen Vorbereitung, Durchführung und Auswertung von Tests für FORTRAN-Programme. KfK-Ext. 13/77-2.

Anderson, T. (1977). Software Fault-Tolerance: A System Supporting Fault-Tolerant Software. INFOTECH State of the Art Conference on Reliable Software, London.

Avizienis, A., and L. Chen (1977). On the Implementation of N-Version Programming for Software Fault-Tolerance during Program Execution. Compsac, Chicago.

Boehm, B.W. (1974). Some Steps toward Formal and Automated Aids to Software Requirements Analysis and Design. 2nd IFIP Congress, Stockholm.

Fischler, M. A., D. Firschein, and L. W. Drew (1975). Distinct Software: An Approach To Reliable Computing. 2nd USA-Japan Computer Conference Tokyo.

Geiger, W., L. Gmeiner, H. Trauboth, and U. Voges (1979). Program Testing Techniques For Nuclear Reactor Protection Systems. Computer, 12, Number 8, 10-18.

Gmeiner, L. (1978). Projektbegleitende Fehleraufzeichnung und -Auswertung während der BESSY-Pilotimplementierung. (unpublished).

IFTRAN: Structured Programming Preprocessors for FORTRAN. (1976). General Research Corp., Santa Barbara.

Jensen, K., and W. Wirth (1974). PASCAL - User Manual and Report. Lecture Notes in Computer Science, Springer, 18.

Jüngst, U. (1976). Design Features of the Fuel Element Computerized Protection System BESSY. IAEA/NPPCI Specialists' Meeting, München.

KTA-Regel 3501 (1977). Reaktorschutzsysteme und Überwachung von Sicherheitseinrichtungen. Kerntechnischer Ausschuß, Fassung 3/77.

PHI2 (1977). Programmierhilfe-Makros für strukturierte Programmierung. SIEMENS Programmbeschreibung P71100-J1015-X-X-35.

RXVP, FORTRAN automated Verification System. (1975). General Research Corp., Santa Barbara.

Voges, U., and W. Ehrenberger (1975). Vorschläge zu Programmierrichtlinien für ein Reaktorschutzrechnersystem. KfK-Ext. 13/75-2, Kernforschungszentrum Karlsruhe.

DISCUSSION

Lauber: Could you explain what is meant by 'specification related errors'?

Gmeiner: The origins of 'specification related errors' are in the program specification phase.

Maki: Has there been the possibility of collaboration between the different teams?

Gemeiner: The teams have been located in the same house. Nevertheless we believe that there has been not very much collaboration and intercommunication. This can also be seen from the very diverse programs produced by the teams.

Bologna: In your presentation you have used the word 'specification' very extensively. But the word 'specification' alone is not very meaningful. Can you explain if you are speaking about 'software-requirements-specification' or about 'program-specifications'? And the errors which you have discovered, were they in the program-specification or in the software-requirement-specification?

Gmeiner: The errors were in the program-specifications. We received the requirements-specifications in natural language spread over a lot of documents. From these documents, we constructed a single document: The program-specifications in a formal manner.

Lauber: I had the same question as Mr. Bologna. The terms 'specification' and furthermore 'software-requirements-specification' are not clear at all. Now some remarks to the way as you use the term 'diversity'. If you take the definition of this term, which was given by Dr. Aitken yesterday, diversity in your system is not sufficiently used. In the words of Dr. Aitken one could say that you specified: 'cross the channel from Dover to Calais by bicycle, and use different types of bicycles', but not: 'cross the channel either by bicycle or by balloon'. But now another question: How did you implement this diverse software? Did you implement it on a 3-computer-system? And my next question would be: how to cope with software-poisoning in this case of diverse software?

Gmeiner: As we had only one computer, we implemented the system on one computer and simulated the 2-out-of-3-system. We ran the three diverse programs one after another and stopped them at internal checkpoints and compared the results. Concerning your next question: We didn't take into account the aspect of software-poisoning.

Schwier: You said diversity will lower the costs of verification. I think, diversity will nearly double both the cost of programming and of verification. If you have two independent teams, you only will lower the costs if the teams are of half size or if you spend half the manhours for each team, compared to equal programming. Then however, the numbers of remaining errors will increase in an unknown manner.
Diverse programming will only have a positive effect on safety if each of the programs was written and verified nearly as carefully as in the case of equal programming.

Gmeiner: It is clear, that you have twice the software production cost, but I don't think that you have twice the costs in the verification area too. You must not test your programs to such a degree as you do if you have identical software in each of the three computers. So your single programs may have lower reliability figures. The necessary high system reliability is achieved by a 2-out-of-3-system by combining less reliable subsystems.

ON A DIVERSIFIED PARALLEL MICROCOMPUTER SYSTEM

R. Konakovsky

*Department of Electrical Engineering and Computer Science,
Northwestern University, Evanston, Illinois 60201, USA*

Abstract. In this paper three different structures of microcomputer systems consisting of two parallel diversified hardware units are described. The use of diversified hardware units implies the use of diverse operating systems, compilers, etc. By comparison of the intermediate results obtained from both hardware units executing the same application software, hardware design errors, errors in operating systems, in compilers, etc. can be detected. Such failure detection capability is the goal for the described system architecture.

Keywords. Parallel microcomputer system; functional diversity; hardware failure defection; software error defection; design error defection; fail-safe microcomputer system.

INTRODUCTION

The design of computer systems, which should meet high reliability or safety requirements is characterized by the use of a number of different methods and techniques, each contributing to failure avoidance, detection, recovery or initialization and carrying-out safety measures. There are many methods described in the literature. Some of them are related only or essential to the software (Avizienis, 1977a; Ehrenberger, 1978; Engler, 1977; Kapp and Daum, 1979; Lauber, 1975, 1978; Taylor, 1976), other to the hardware (Santucci, 1977; Schneeweiss, 1978; Schüller and Schwier, 1977; Strelow and Vebel, 1978). There are also methods related to both, hardware and software systems (Avizienis, 1977b; Ciompi and Simoncini, 1977; Konakovsky, 1975, 1977, 1978; Maki, 1978; Nilsson, 1978; Plögert and Schüller, 1977; Rennels, Avizienis and Ercegovac, 1978; Sedmak and Liebergot, 1978; Trauboth, 1975; Wobig, 1978).

Parallel redundancy with comparison is a well known technique for failure detection. If one of the parallel units fails, it can be detected by the comparator. There are some limitations to its failure detection capability. First, common and multiple failures can remain undetectable. To this category belong e.g. logical or design errors. Second, the comparator can fail in some undetecable way.

Many different techniques have been described in the literature to reduce the probability of undetectable failures of both categories. For instance to reduce double failures in functional logic of VLSI chips, a duplicate complementary logic using complementary signals is proposed as a parallel redundant unit (Sedmak and Liebergot, 1978). For this complementary logic a separate design and a new mask is necessary to implement it on a chip. Self-checking digital network is used to implement a comparator (Sedmak and Liebergot, 1968).

Another technique for considerable reduction of double failures applies dynamic switching between original and complementary logic of both parallel redundant units (Lohmann, 1979). The comparison of all original signals with their complementary ones is carried out synchronously on the gate level by means of a special comparison circuit.

To detect failures in distributed computer systems special design technique is proposed by Rennels, Avizienis and Ercegovac (1978). The basic architectural unit is a small computer (e.g. microcomputer), which is capable of detecting internal faults during normal operation. This so-called "self-checking computer module" uses e.g. two identical CPU's to detect fault operations by a disagreement comparison.

A special design of a microcomputer system called SIMIS (Sicheres Mikrocomputersystem) is described by Strelow and Vebel (1978). Two identical microcomputers are used in parallel synchronous mode of operation in this system. Failures are detected by the disagreement comparison of their bus signals at each clock period.

In this paper the design of a microcomputer system using two parallel diversified hardware units is discussed. This method contributes to failure detection in system hardware and software. The use of parallel diversified hardware units allows during the system operation (on-line) the detection of almost all possible failures of the executing system. This is important when recovery or safety operations have to take place before damages to the system or to persons can appear as a consequence of failures.

Although errors in the application software are not considered by this approach, they can be detected applying some methods described in the literature, e.g. methods of funkctional diversity (Engler, 1977; Lauber 1975, 1978), or N-version software (Avizienis, 1977a). This is discussed at the end of the paper.

BASIC STRATEGY

The diversified microcomputer (DIMI) system uses two parallel diversified hardware units and a comparator. The simplest structure is shown in Fig. 1.

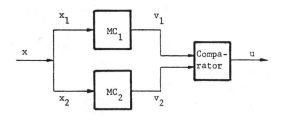

Fig. 1. The simplest DIMI structure

MC_1 and MC_2 represent two different (diverse) microcomputers.
The output signals v_1 and v_2 are compared by the comparator. In the case of agreement the output u of the comparator corresponds to v_1 and v_2. Otherwise recovery or safety measures must be initiated by the comparator. There are some problems associated with this structure.

The use of diversified hardware units requires data conversion to a common format before output signals can be compared. This can be carried out by the comparator. Such system requires a complex comparator.

Another problem is the comparison efficiency. In order to achieve shortest possible error detection time, the comparison must be as often as possible and it should include as many signals as possible. The use of diversified hardware units makes the comparison on the gate level impossible. The application software level only is convenient for any comparison. One of the most frequent comparisons, which can be achieved for an application software currently executed, is the comparison of all intermediate results currently obtained as well as combined with the periodical comparison of all variables used by this software. All variables on the left side of any assignment statements (characterized by:=) can be considered as intermediate results when currently executed. For instance, in the following program written in PEARL-language, the values of I an A (I) of the first, I and B (I) of the second and ALPHA of the third assignment statement are intermediate results to be compared when executed by both hardware units.

```
FOR I FROM 1 BY 1 TO N
REPEAT
A(I) :=   C(I) * SIN (ALPHA);
B(I) :=   C(I) * TAN (ALPHA);
ALPHA:= ALPHA + K * DELTA;
END;
```

Besides of the comparison of varibles currently obtained the periodic comparison of all variables used by the application software is possible. The time for such comparison must be chosen in an appropriate way, when the values of variables to be compared correspond to their updated values in both hardware units. A comparison of old values from one hardware unit with new values from the other hardware unit as a consequence of asynchronnous mode of operation, is to be avoided. This can be achieved, for example, if the time for such comparison is determined by a state when both microcomputers are waiting for an interrupt signal from the process. Such states are very often in real-time applications. Instead of background program execution, the desired comparison can take place.

The comparison itself is not without problems. As mentioned above data conversion to a common format is necessary. If integer values are to be compared, such conversion is without difficulties. Problems can arise if real values are to be compared. A comparison with respect to a given accuracy (tolerance) has to be applied. This implies greater complexity of the comparison unit.

Another problem arises from the use of real (floating point) numbers. Consider an input variable, e.g. reactor temperature (variable TEMP) is represented by a floating point number in both hardware units. A conditional statement, such as

```
IF TEMP >100.00 THEN ALARM :=1;
                ELSE ALARM :=0;
FIN;
```

can result in different branching, e.g. if the values in the parallel hardware units are 100.000 and 100.0001 respectively. To force the same execution of the application software in diversified hardware environments all branching statements, which depend on the tests with real numbers, must be evaluated with the same given accuracy in both hardware units. If the given accuracy for our example above is one decimal digit behind the decimal point, the same branch result, i.e. ALARM :=0; is then obtained. This implies additional requirements to the compilers.

As shown above, the comparator for diversified hardware units is much more complex than for identical hardware units. A comparison on the gate level with self-checking digital network as described by Sedmak and Liebergot (1978) is not possible. The implementation of the comparison unit by any special, but non LSI or VLSI network like URTL-System (Lohmann, 1979), seems not to be meaningful. The use of two diversified comparators implemented by LSI or VLSI building blocks seems to be a good solution.

Fig. 2a) shows the block diagram of the diversified microcomputer (DIMI) system with two comparators C_1 and C_2. In the DIMI system each input variable x is represented by two input signals x_1 and x_2. Similarly, any output variable u is represented by two signals u_1 and u_2 obtained from the two diversified hardware units. Thus any output variable can be used as an input variable for another DIMI system. The current output signals v_1 and v_2 of the microcomputer MC_1 and MC_2 as well as the values of all their internal variables w_1 and w_2 are compared by two diversified comparators C_1 and C_2. Their outputs u_1 and u_2 correspond to the values v_1 and v_2, respectively, in the case of agreement at both comparison units. In order to meet safety requirements the proper function of both comparison units must be checked within an appropriate time period. This can be provided applying some methods known from the literature (Maki, 1978; Nilsson, 1978).

Fig. 2b) shows the block symbol for the DIMI system. The double lined input and output connections symbolize the use of two signals for each variable.

IMPLEMENTATION

The above discussed basic strategy of the DIMI system resulting in the block diagram of Fig. 2a) can now be implemented in different ways. Three structures of possible implementations are shown and described in more detail. In all three cases the comparison units C_1 and C_2 are implemented by the same building blocks as microcomputers MC_1 and MC_2 respectively. Thus, they possess the same hardware diversity.

Fig. 3 shows the first implementation structure. The comparison units C_1 and C_2 are realized by physically separate microcomputers of the same type as MC_1 and MC_2 respectively. The main difference between the comparators (C_1, C_2) and the microcomputers (MC_1, MC_2) is the size of the memory used (RAM and ROM size). The comparators use only a small memory to realize the comparison function. In order to compare all intermediate results (variables w_1 and w_2) they have to appear on the output of the microcomputers MC_1 and MC_2 respectively. Specified output addresses in O_1 and O_2 can be dedicated for the output of intermediate results. During the compilation of the application software the compiler must insert an output statement after each assignment statement.

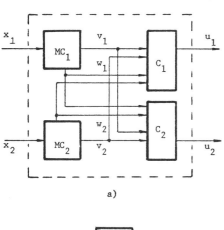

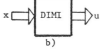

Fig. 2. Block diagram of the DIMI system

Example:

```
A(I) := C(I) * SIN (ALPHA);          ─┐
W(K) := A(I);                         ├ statements inserted by
IF   K< 15 THEN K := K+1 ELSE K := 0; ─┘ the compiler during
B(I) := C(I) * TAN (ALPHA);          ─┐  compilation
W(K) := B(I)                          │
IF   K< 15 THEN K := K+1 ELSE K := 0; ─┘
```

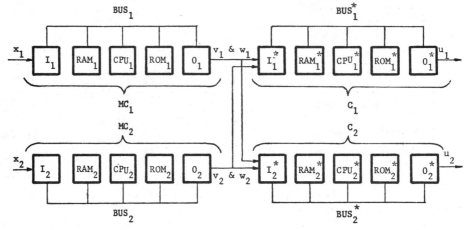

Fig. 3. DIMI System of the "4CPU/4BUS" type

In this example the output array W(0:15) is dedicated for the comparison of intermediate results. The IF statement implements the increment of the Index K modulo 16.

The insertion of statements necessary for the vomparison function can also easily be implemented by a microprogram. For example, two different machine instructions for data store can be distinguished: with and without an additional data transfer for comparison. The first instruction can be interpreted directly by the microprogram in the desired way without the necessity of any statement insertion by the compiler.

The comparison units must be capable of suspending one or both microcomputers if the output array W is full.

On the other hand, when one or more comparisons of W are accomplished and the output array W has free spaces available, the comparison units must be capable of restarting the suspended microcomputer. For this purpose interrupt signals can be used. These signals are not shown in Fig. 3 the output signals v_1 and v_2 of the microcomputers MC_1 and MC_2 appear as the output signals u_1 and u_2 of the DIMI system after the agreement comparison by the C_1 and C_2 units. The described implementation structure is characterized by the four physically separate units MC_1, MC_2, C_1 and C_2 respectively. It can be classified as a "4 CPU/ 4 BUS" type DIMI system.

Another implementation structure is shown in Fig. 4

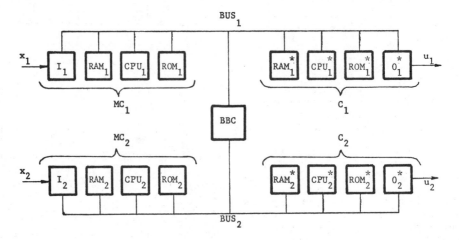

Fig. 4. DIMI System of the "4 CPU/2 BUS" type

The four buses are merged into two buses, and the cross-communication between microcomputers and comparators is provided by the bus-to-bus communication (BBC) unit. In the simplest way this BBC unit can be implemented by the input/output units as shown in Fig. 5.

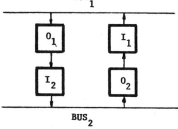

Fig. 5. Bus-to-Bus Communication with I/O units.

The two microprocessors in MC_1/C_1 (MC_2/C_2) units share one single bus. In order to assure that both processors can access different memory modules in one memory cycle time, the bus speed must be twice as fast as that of the memory. In this case both processors sharing one single bus can operate in parallel without a speed penalty. Under this condition the DIMI system in Fig. 4 has the same computation capability as in Fig. 3. The important differences, however, are in the reliability or safety figures which are chievable by these structures.

The system in Fig. 4 can continue its operation when one or even two processors fail (fault-tolerancy). The remaining processor (on one or on both buses) has to carry out both functions: that of the microcomputer and that of the comparator as well. This improves the reliability considerably. A certain degradation e.g. lower computation speed can appear. On the other hand, instead of 4 physically units, there are now only two or one; it depends on the implementation of the bus-to-bus communication (BBC) unit. This can have an adverse effect to the safety. Although the physical separation of the units is not so clear as in Fig. 3, there is still a physical separation of functions using separate microprocessors for this implementation. A proper design of this system is necessary in order to meet safety requirements.

The above discussion about fault-tolerancy gives an idea of the third DIMI structure using only one microprocessor at each bus for both functions; that of the application software and that of the comparison unit. This structure differs from Fig. 4 only in the missing of two CPU's one at each bus. No other unit can be saved. The RAM and ROM modules of the saved microprocessor are now used by the other microprocessor, in order to realize the desired function. The price to pay for is time. The third structure is of a "2CPU/2 BUS" type and can be considerably slower than the "4 CPU/2 BUS" type.

DIMI-NETWORK

The above described diversified parallel microcomputer (DIMI) system can be used to design a complex network. The DIMI system is then used as a module. In such a network each variable to be transfered from one DIMI module to another is represented by two signals. The DIMI modules can be connected in two ways:
a) Input-Output connections between two DIMI modules are from the same (non-diversified) hardware. No data conversion is necessary.

b) Input-Output connections between two DIMI modules are from the diversified hardware. Data conversion can be required.

To assure proper transmission, all input variables can be tested before their use, by the comparison of their two signal representations.

Fig. 6. shows a simple example of a DIMI network used for process control. The block symbol from Fig. 2b) is used for DIMI modules.

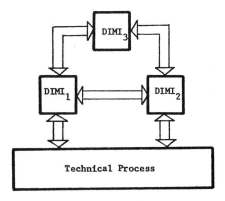

Fig. 6. DIMI - Network for process control (example)

The double line connection symbolizes the double signal representation of all variables. The DIMI-Network can be viewed as a distributed computing system. The failure detection capability of the DIMI module can be used to enhance the reliability or safety of the entire system. If one DIMI module fails, it may cause a graceful degradation of the system, but does not necessarily lead to a total breakdown. The safety of the entire system can stay unaffected. To achieve the above discussed reliability or safety characteristics of the DIMI network an appropriate software system is necessary. This Problem is briefly discussed in the next section.

RELIABILITY AND SAFETY CONSIDERATIONS

In order to meet the high reliability or safety requirements of the entire system, both system parts, the DIMI modules as well as the application software modules, have to meet these requirements.

The use of diversified parallel hardware units in the DIMI module should assure that almost all possible failures are detectable. Such failure detection capability can then be used to meet the reliability and safety requirements of the entire DIMI network.

For the application software modules, which reside in DIMI modules, two different methods can be used:

a) Perfectionist approach

b) Functional diversity approach

The first method can be used only for a software of a small size. This approach should produce error-free software modules.

The second method should produce diversified software modules. The functional diversity of the modules should exclude the possibility of the same error in these diversified modules. Two DIMI modules can be used for the execution of two diversified application software modules. The comparison of their output values as well as intermediate results should guarantee the detection of all possible errors in both software modules. Such error detection capability can then be used to meet the reliability and safety requirements of the entire application software system. This brief discussion should show that DIMI modules can be used to design a complex network, which must meet high reliability or safety requirements.

CONCLUSION

In this paper a diversified parallel microcomputer (DIMI) system is described. Some advantages of the DIMI system, such as the capability to detect hardware design errors, compiler errors, errors in the operating system or errors arising from hardware-software interactions, can be decisive for the use in a system, where high reliability or safety are required. To achieve this failure detection capability the DIMI system must be designed carefully. This includes also the choice for appropriate microprocessors, RAM or ROM chips. The use of microprogrammed microprocessors can be useful but may involve some new problems, such as the necessity of diversified microprograms. Although a careful design is important, there are no special restrictions in the choice of hardware building blocks, exept of the diversity. This is impotant when DIMI systems should be redesigned using new technology.

ACKNOWLEDGEMENT

This work was supported by the Deutsche Forschungsgemeinschaft, Germany. The autor wishes to thank Professor S. S. Yau, Northwestern University, Evanston, Illinois, USA, and Professor R. Lauber, University of Stuttgart, Germany, for many helpful suggestions.

REFERENCES

Avizienis, A. and Chen, L. (1977a). On the Implementation of N-Version Programming for Software Fault-Tolerance During Program Execution. Proceedings of the COMPSAC Conference, Chicago, pp 149-155.

Avizienis, A. (1977b).Fault-Tolerant Computing-Progress, Problems and Prospects. Proceedings of the IFIP Conference, pp 59-72.

Ciompi, P. and Simoncini, L. (1977). Design of Self-Diagnosable Minicomputers Using Bit Sliced Microprocessors. Design Automation & FTC, Vol 1, No 4, pp 363-375.

Ehrenberger, W. D. (1978). TC 7 Working Papers on Software Reliability since 1976.Purdue Europe TC 7 Safety and Security, Working Paper No 162.

Engler, F. (1977). Safe Software Strategies. Purdue Europe TC 7 Safety and Security Working Paper No 133.

Kapp, K. und Daum, R. (1979). Sicherheit und Zuverlässigkeit von Automatisierungssoftware. Informatik-Spektrum 2,S. 25-36.

Konakovsky, R. (1975). New Method to Investigate the Safety of Control Systems. Proceedings of the IFAC-Congress in Boston/USA, August 1975, paper No. 26.5

Konakovsky, R. (1978). Safety Evaluation of Computer Hardware and Software. Proceedings of the COMPSAC Conference, Chicago, pp 559-564.

Lauber, R. (1978). Software Strategies for Safety Related Computer Systems. Purdue Europe TC 7 Safety and Security, WP No. 163.

Lauber, R. (1975). Safe Software Strategies by Functional Diversity. Purdue Europe TC 7 Safety and Security, Working Paper No. 37.

Lohmann, H. Z. (1979). URTL Schaltkreissystem U1 mit hoher Sicherheit und automatischer Fehlerdiagnose SIEMENS Zeitschrift 48, H7, S. 490-494.

Maki, G. K. (1978). A Self Checking Microprocessor Design. Design Automation & FTC, Vol. 2, No 1, pp 15-27.

Nilsson, S. A. (1978) Selbsttestverfahren für Mikrorechner. VDI-Berichte Nr. 328, S. 77-85.

Plögert, K. and Schüller, H. (1977). Process Control with High Reliability; Data Dependent Failure Detection Versus Test Programs. Purdue Europe TC 7 Safety and Security, Working Paper No. 139.

Rennels, D. A., Avizienis, A. and Ercegovac, M. (1978). A Study of Standard Building Blocks for the Design of Fault-Tolerant Distributed Computer Systems. Proceedings FTCS-8, Toulouse, pp 144-149.

Santucci, C. (1978). Safe Hardware - Details of Checking Programs. Purdue Europe TC 7 Safety and Security, Working Paper No. 144.

Schneeweiss, W. (1978). Methoden zum Aufbau zuverlässiger und sicherer Systeme. VDI-Berichte Nr. 328, S. 13-16.

Schüller, H. and Schwier, W. Safe Computer Systems - Hardware. Purdue Europe TC F Safety and Security, Working Paper No. 97.

Sedmak, R. M. and Liebergot, H. L. (1978). Fault- Tolerance of a General Purpose Computer Implemented by Very Large Scale Integration. Proceedings FTCS-8, Toulouse, pp 137-143.

Strelow, H., Vebel, H. (1978). Das Sichere Mikrocomputersystem SIMIS. Signal & Draht 70, 4 S. 82-86.

Taylor, J. R. (1976). Strategies for Improving Software System Reliability. Purdue Europe TC 7 Safety and Security, Working Paper No. 48.

Trauboth, H. (1975). Methods to Develop Safe Computer Systems. Purdue Europe TC 7 Safety and Security, Working Paper No. 35.

Wobig, K.-H. (1978). Sichere Systeme und deren Grundlagen in der Eisenbahnsignaltechnik. VDI-Berichte. Nr. 328, S. 87-91.

DISCUSSION

Trauboth: Could you explain the difference between 'safe application software' and 'safe execution of the user-software'?

Konakovsky: The application software means a program written by a user for example in a high programming language. In order to make it run you need to compile it to the machine code, you need an operating system and a hardware system as well. I understand the execution system as everything necessary to make application software operational. Safe execution of the user software means that there is no error or failure in the execution system, i.e. in the compiler, operating system and in the hardware which could affect the safety adversely. On the other hand safe application software means that such an error does not contain the application software. To assure this in both parts of the system, in the application software and in the execution system, the principle of diversity must be applied to both parts independently.

Parr: I've got two or three problems with what you said. The first one is that in order to do your synchronization you must check intermediate values. First of all this seems to me to assume one set of application software, which implicitly prohibits the use of diverse application software. And secondly in order to do the synchronization, because I can imagine the situation that you have two independent operating systems and so on, then if one piece of software runs slightly faster than the other, you are going to get the situation that the triggers to the two computers will be different, so they will be ready to see the interrupt in the one processor and not in the other and the actual cause of computation may be different. In order to prevent this you are going to need some software synchronization. And this synchronization will provide another common failure mode.

Konakovsky: To the first question, to the use of diverse application software: This system is proposed in order to assure that all errors and failures in the execution system, because of its diversity, are detectable. You can use two such systems to run two diverse application software. In this case each diverse part is executed twice, for example the application software number one is executed twice and the application software number two is also executed twice, each one by two diverse execution systems. A comparison between two diverse application software is possible and must be solved by the user which results to safe application software. This means that the proposed system can also be used for diverse application software.

To the other question about the problem of synchronization: You are right, you have to solve the synchronization problem. This must be solved in both parts separately, in the application software and in the execution system as well.

In order to avoid common failure mode in the synchronization of diverse parallel systems a proper design of that small part of the system which cannot be diversified (hardware and software) must be assured.

Trauboth: Related to synchronization: Does it mean you need also more checkpoints in order to prevent all the error propagation. Wouldn't that slow down the operation?

Konakovsky: No. Synchronization causes that the computation speed is according to the slower system. The comparison between the systems is performed in parallel. Therefore the number od checkpoint has no or only a negliable impact to the computation speed.

Trauboth: But you can have these synchronizations on an instruction level or on the module level. You would have checkpoints where you have to stop the process and see whether the other process is in time or not.
What level of detail is the optimum?

Konakovsky: The optimum depends on the particular implementation.
One possibility to solve the problem of synchronization is to store intermediate results in an array, which serves as a buffer. When the array is full a comparison with the corresponding array of the other process must take place before new intermediate results can be stored.

Maki: What are your applications suitable for?

Konakovsky: The proposed system can be used in any application where safety in addition to reliability is required, such as in railway automation. The system is suitable for example as a module in a distributed computing system which has to meet safety requirements.

AN INVESTIGATION OF METHODS FOR PRODUCTION AND VERIFICATION OF HIGHLY RELIABLE SOFTWARE

G. Dahll and J. Lahti*

OECD Halden Reactor Project, Halden, Norway
*Technical Research Centre of Finland, Espoo, Finland

Abstract. A joint project between Technical Research Centre of Finland and OECD Halden Reactor Project with the aim of investigating methods for developing and testing computer programmes with high requirements for reliability has been completed. This paper gives a short outline of the project. The main emphasis is on the experiences gained during the project and on the conclusions and recommendations made on the basis of this experience. Methods used for the specification, programming, programme analysis, testing, and reliability assessment are discussed together with their virtues and weaknesses.

Keywords. Computer software; structured programming; quality control; software reliability; system analysis.

INTRODUCTION

A joint project between the Technical Research Centre of Finland (VTT) and the OECD Halden Reactor Project (HP) on investigations of methods for production and verification of computer software with very high reliability requirements has now been completed. In the following a short description of the project is given, emphasizing the conclusions and recommendations based on the experience from the project.

The intention of the project was to investigate the applicability of various methods for producing and testing high reliability software in order to gain experience with such methods. Due to the limited resources available, the approach was to cover a fairly broad area rather than going in depth on any single method. However, the experiences gained should be of value in identifying areas where further research is appropriate.

The work was concentrated on the reliability of safety-related software. This implies that the programme is fairly simple, has very high reliability requirements and in a real application on a plant probably be implemented on a dedicated computer. However, many of the results from the project may also be of value for production and verification of other types of computer programmes.

The project started with the selection of a typical safety-related problem in a nuclear reactor which is natural to solve by a computer, viz., a calculation of the departure from nucleate boiling ratio, combined with various types of consistency checks on the input data. A complex input data checking routine was added to the programme in order to obtain experience in the programming and analysis of complex logic software.

The project had five phases: specification, dual programming, programme analysis, programme testing, and reliability assessment. Part of the work was performed jointly by the VTT and HP staffs and portions were done in parallel with each staff working independently.

SPECIFICATION

The specification was produced as a joint effort and written in plain English, in a form we thought it would have been written by a reactor engineer. The reason for this was that we wanted to determine whether there were points in the specifications that would be interpreted differently by independent teams of programmers. The objective was to identify the characteristics needed to make specifications clear and uniquely understandable by a programmer.

In the specification effort special emphasis was put on preventing programme errors due to incompleteness in the specifications. Our intention was to define precisely what the programme should do for all combinations of input data. Validity areas for all input data were defined, as well as the validity area of the arguments of the formulas to be used in the programme. Control strategies were specified to check these criteria. Despite these efforts, the specifications contained many incomplete points and some contradictions which were revealed during the programming and testing phases. A typical example of incompleteness, involved one of the data checks which compared the value of one process variable with the value of the same variable in the previous cycle and checked for a drastic change. The specifications failed to define how this check was to be performed for either the first cycle or for the case where value from the previous cycle had been rejected. Most of the specification errors were related to situations of this type, which can be classified as "programme startup" and "rejected input processing".

During the work on programme development it became evident, as a first conclusion from the project, that the specification phase was the weak link in the programme development chain. We think this is partly caused by the lack of a formal specification method, and recommend that such a method with the following characteristics should be developed:

- The specification should be clearly understood by both the engineer and the programmer because it is the principle link between them.
- A clear syntax for the specification language must be defined. The methods of top-down structured programming should be used also in the specification phase, for three reasons:
The person making the specification is forced to attack the problem in a systematic way, the resulting specification will be made in a form which eases the programmer's work in transforming it into a structured programme, and the comparison and verification of the final programme with the specification is made significantly simpler.
- The specification method should also include a procedure for the specification. This would aid the specifyer in the problem analysis by forcing him to consider all sides of the problem. An interactive computer programme might be advantageous, as such a programme may point out sides of the problem which are still not specified, and further check the completeness and consistency of the specifications.

PROGRAMMING

One principal experiment in the project was the use of dual programming in the programming phase. Both teams, working in parallel and independently, made their own version of the programme using their own programming methods.

During the programming phase both teams used the procedure of letting an inspector supervise the programmer's work. It is the conclusion of both teams that this method is advantageous because many programming errors are found at an early stage, and also because the programmer has to formulate his algorithms so clearly that the inspector can understand them. It became evident, however, that the inspector's work was influenced by his reading of the programmer's descriptions. To cope with this, the inspector should apply some systematic methods for programme analysis. It is therefore recommended that the programme development should follow some strict rules like top-down structured programming, description of how variables are handled by each module, description of the data flow in the programme and the entry and exit conditions for each module.

The Design

It was attempted to develop the programme using a general design form. This form should be easy to understand for the programme analyst and should have features easing the analyst's work. It should also be independent of coding language and easy to translate into any high level language. However, the two teams used different design methods: structured flow graphs and pseudo-code.

Structured Flow Graphs. With this method the design is made with a special type of diagrams. These diagrams resemble flow diagrams in their appearance, but are used as structured flow charts (Nassi and Schneidermann, 1973). Some characteristics of the method are:

- The programme is built up in a top-down fasion so that a diagram represents one programme module which can contain sub-modules at an lower level. A module can only have one entry and one exit.
- One module is described on one sheet of paper. Off-page connectors are not allowed.
- Only three constructions are allowed: Sequence, while-do and if-then-else.

For each module there is a heading containing a description of what the module does and a list of all identifiers used by the module. The identifiers are classified in four categories according to how they are used by the module: defined, used, local variables, and local constants. The classification of the variables is helpful in the construction of the programme in a top-down modular way. The external variables of one module constitute the connection to the encompassing module, whereas the local variables are known only inside the module. Variables should be kept as local as possible, thus allowing various modules to be relatively independent of each other. Further, this makes it easier to understand the data flow within one module, without studying the details of the submodules. All of these features are useful in the programme analysis phase.

The flow graph method has the advantage that it makes the control flow of the programme easy to grasp visually. This eases the programme analysis. The drawback of the method is that it must be drawn manually, a shortcoming which became particularly evident when it was necessary to make corrections. In the construction of flow graphs one should be careful to choose the module sizes that produce a comprehensive programme. The modules should neither be too large or too small. In this project the module sizes used proved to be smaller than desirable and an excessive number of modules were produced.

Pseudo-code. Pseudo-code is a design language which resembles a computer programme in its control structure, but where one is free to use ordinary language in the description of details. The pseudo-code should be clear enough to avoid conflicts in the interpretations. It should also be capable of representing the control structures of structured programming. The pseudo-code was selected to have Algol-like control structures and statements. During the programme design, the structure of the pseudo-code was changed only in case of a failure so that most of the difference between the design steps was only in the degree of details.

The following is an example of a piece of pseudo-code

in an early design phase:

> if (measurement k is not rejected)
> then if (the difference between measurement k and the estimated group value $\phi_{i\ JREJG}^{a}$ is less than NEAR)
> then begin 4 this measurement k is the nearest sofar
> NEAR = the difference between the estimate and measurement k
> end 4

The declarations and initializations were not made in the pseudo-code because of the amount of handwriting. This can be considered as a mistake, as there were some errors in the coding phase which could have been avoided if initializations and declarations had been present already in the design phase.

The pseudo-code does not have the visual power of the flow graphs. It has, however, the advantage that it can be written with plain letters. Although it was not done on this project, it is clearly feasible to store the design in a computer and use a simple editing programme to make corrections. Through a simple correspondence between the pseudo-code and the final programme, it should also be possible to make a computer programme that would check the consistency of the final programme and the initial design.

The Coding and Implementation

The coding was made in two different programming languages: flow-graphs were converted to PASCAL and pseudo-code was converted to FORTRAN. Neither language was ideal for modular programming because the modularity had to be implemented by the use of sub-routines in both programmes. A feature of the PASCAL language which should be used extensively, is the possibility of defining data types. A sensible use of this feature makes the programme easier to read and correct, which in conjunction with run-time checking should increase the reliability of the final programme.

One general conclusion we can make on the basis of our experience with the programme development, is that one should develop some form of computerized and to assist in the process. There are two obvious areas where a computer programme could be used to increase the software reliability. One is to write the various steps in the programme development process in a form so they can be stored in a computer. A computer programme could then be used to check the consistency of each step with the previous ones. The other area would employ a computer to perform various consistency and completeness checks on the modules of the programme.

PROGRAMME ANALYSIS

After completion of the programming phase, the two teams seeded a set of artificial errors into their own programmes and exchanged them for a mutual programme analysis. The intention was to investigate the applicability of various analysis methods and their ability to detect programme errors. One of our conclusions is that the more elaborate methods should mainly be used to verify correct programmes and not as a debugging device. The various analysis methods should be performed in the order of increasing complexity. One should follow a pre-defined schedule starting with straight-forward checking of declarations, array bounds, parameter lists, etc., continue with data flow and control flow analysis, and ending, if desirable, with the more formal programme proofs. The programmer should also prepare the programme for analysis. This should be done by documentation of the programme modules by functional descriptions, description of all variables, data flow description and definition of entry and exit conditions for all modules.

Both teams used in principle the same methods for the programme analysis: First the programmes were divided in modules in a top-down fashion. These modules were not necessarily of the same size as the one made by the programmer, but rather in sizes which were a suitable unit for analysis. These modules were written in a diagrammatic form like the structured flow graphs, and the variables in each module were classified. This classification was used in a tabular method to check the consistency in the data flow. The VTT team also derived the entry and exit conditions for each module in the form of logical expressions between the values of variables. These conditions were used to detect contradictions and incompleteness in the code. Some attempts were also made to carry out formal programme proofs for some of the modules. However, formal programme proofs are very laborious to make manually, and are also error prone. It is our conclusion that such proofs are best suited for handling small programme modules with complex logical structures. There is also the risk that in the manual execution of a correctness proof that the analyst's interpretation of the modules functions may be incorrect. An example of this type in fact occurred. Therefore automatic aids which can analyse the programme source code should be used for formal programme proof if they are available.

In our project all of the programme analyst's had been involved in the independent programming of a different version of the programme they were analyzing. This made their analysis more independent of the programme they were analyzing and also made it easier to reveal ambiguities in the programme specifications. Our experience also leads us to the conclusion that the method of introducing artificial errors is useful in testing the efficiency of the analysis. The errors should be well hidden and scattered throughout the programme. They should, however, be easy to correct and not so significant that they will confuse the analyst's understanding of the programme.

A general conclusion of both teams is that the methods of programme analysis should be applied in an as early stage of the programme design as possible in order to prevent fatal influences from early design errors. A change in the programme always involves more effort when an error is found in a late phase and the risk of introducing an uncorrected error in the revision is also larger.

TESTING

One of the main advantages of dual programming is that one can make simultaneous test runs of the two programmes. This is a powerful method for testing identical performance of the two programmes, and as good a test as any of the correctness of the programme compared to their specifications. There are various reasons why this testing method is advantageous. First, the method makes it possible to make an extensive test with a large number of test data within a reasonable time. Second, it is the compiled programme implemented in a computer which is tested. In this way errors due to misconceptions of the programming language, compiler errors and implementation errors will be revealed, in addition to programming errors. However, to make a more complete test with significant results, the test should be made on the actual computer on which the programme should run, and under realistic timing conditions and with a realistic way of reading and writing data. Furthermore, such a test makes it possible to test all aspects of the programme performance under normal as well as anomalous conditions. This will often reveal ambiguities in the specifications, since such ambiguities are often connected to anomalous situations.

The significance of the results of the test runs is dependent on the selection of test data. It is important to select data so that all aspects of the specification are tested as well as all parts of the programme. To fulfill this condition and also to prevent systematic errors, we used a certain degree of randomness in the selection of data.

Three kinds of simultaneous tests were performed:

- A so-called sequential random test, where the test data constituted a set of consecutive data forming a series in time. The time series were constructed from Gaussian distributions and caused all segments of the FORTRAN programme to execute during the test. The effectiveness of the test was enhanced by frequent changes of the mean and the variance of the distributions. The FORTRAN programme was instrumented with counters to ensure the execution of all segments. Tests of this type were not performed on the PASCAL programme, but it is reasonable to assume that all segments were executed also there.
- A start-up test, where the first few cycles of consecutive executions of the programmes were tested. This test has significance because the programmes are time-dependent using the previous input data and some of the results from one computation cycle are used in the next cycle. The start-up was also rather imprecisely specified, so one would expect many errors in that part. The test data generation programme was the same as in the previous testing method and the programmes were automatically reinitialized after a few cycles.
- A process simulation test where the programmes were tested with data that were generated by the simulation of a realistic variation of the flux and steam quality in a reactor channel. This test detected no programming errors, but revealed some weaknesses in the specifications with respect to the dynamics of the real process.

More than 20,000 test cases were executed and altogether 29 new errors were found of which 26 were detected in the first 1,000 cycles. The last 15,000 cycles were executed without detecting any new errors. There were, however, two errors detected during this phase which had been introduced as a result of the correction of earlier errors.

The testing methods were applied also with the artificial errors which were seeded into the programmes for programme analysis. Of these errors all, except two, were found quite early in the test. An investigation of these two cases revealed that neither of them would influence the result of the computation for any combination of input data. The errors in the programmes found during various phases of the programme development are summarized in Table 1.

The reader should note that because of practical reasons the PASCAL-programme was not executed prior to the programme analysis phase. Therefore the number of errors in the different phases of the project are not directly comparable.

RELIABILITY ASSESSMENT

First estimates on the error content of each programme were made on the basis of artificial errors used in the programme analysis phase. The expected number of remaining errors can be easily calculated when the total number of artificial errors, the number of artificial errors found, and the number of original errors found are known (Mills, 1972). The estimates for remaining errors in each programme were 13 and 6. The true numbers found later in the testing phase were 15 and 5, respectively. These numbers fit very well, but the data is too small for making any strict conclusions on the method.

Various methods proposed for quantitative reliability assessment were further applied to the data acquired during the debugging and testing phase. These methods, however, gave diverging results. The methods used were:

- Jelinski-Moranda De-Eutrophication Model
- Jelinski-Moranda Geometric De-Eutrophication Model
- Geometrical Poisson Model
- Schick-Wolverton Model
- Reliability Growth Model
- Weibull Model

The models are described in reference (Moranda, 1975) and (Sukert, 1976). Geometric De-Eutrophication Model and Geometrical Poisson Model gave the best results in the comparison with the actual project results. These are also the only models for which the results are consistent with the decreasing failure rate which one intuitively expects.

The other models are not suitable to handle the failure data from our experiment. In each case the main reason is probably that there are sequences in the failure data where the failure rate is instantaneously increasing.

The Geometric De-Eutrophication model assumes that the hazard rates between successive errors form a geometric series. The hazard rate in the i'th interval between errors is

TABLE 1 Reported Failures in the Programme

	TOTAL				PROGRAMMING PHASE				PROGRAMME ANALYSIS				TESTING PHASE			
	TOTAL		PROT		TRICOT		PROT		TRICOT		PROT		TRICOT		PROT		TRICOT	
	n	%	n	%	n	%	n	%	n	%	n	%	n	%	n	%	n	%
Clerical errors	38	24.7	11	20.8	27	26.7	10	30.3	11	27.5	1	14.3	15	42.9	–	–	1	3.8
Logical errors	36	23.4	8	15.1	28	27.7	8	24.2	10	25.0	–	–	12	34.3	–	–	6	23.1
Ambiguous specifications	18	11.7	4	7.5	14	13.9	–	–	6	15.0	2	28.6	3	8.6	2	15.4	5	19.2
Violation of specifications	14	9.1	4	7.5	10	9.9	1	3.0	6	15.0	2	28.6	4	11.4	1	7.7	–	–
Initialization errors	13	8.4	5	9.4	8	7.9	4	12.1	7	17.5	1	14.3	1	2.9	–	–	–	–
Specification errors	6	3.9	3	5.7	3	3.0	–	–	–	–	1	14.3	–	–	2	15.4	3	11.5
Data transfer errors	6	3.9	6	11.3	–	–	6	18.2	–	–	–	–	–	–	–	–	–	–
Rounding errors	5	3.2	5	9.4	–	–	–	–	–	–	–	–	–	–	5	38.5	–	–
Compiler errors	4	2.6	–	–	4	4.0	–	–	–	–	–	–	–	–	–	–	4	15.4
Syntax errors	4	2.6	4	7.5	–	–	4	12.1	–	–	–	–	–	–	–	–	–	–
Total	144	93.5	50	94.3	94	93.1	33	100.0	40	100.0	7	100.0	35	100.0	10	76.9	19	73.1
Correction errors	10	6.5	3	5.7	7	6.9	–	–	–	–	–	–	–	–	3	23.1	7	26.9
Total	154	100.0	53	100.0	101	100.0	33	100.0	40	100.0	7	100.0	35	100.0	13	100.0	26	100.0

PROT is the FORTRAN programme and TRICOT is the Pascal programme

$$z(t_i) = D \cdot k^{i-1}$$

where D and k can be found as maximum likelihood estimates. The corresponding reliability and mean time to failure are

$$R(t_i) = \exp(-Dk^n t_i) \text{ and}$$

$$MTTF = 1/Dk^n$$

where n is the total number of time intervals. When the test cases between successive errors are used as t_i's the model estimates are given in the Table 2.

TABLE 2 Geometric De-Eutrophication Model Estimates

n	k	D	MTTF (cycles)	actual time to next failure
1	–	–	–	1
2	2.000	0.500	- 0.5	1
3	1.414	0.554	0.6	2
4	1.000	0.667	1.7	10
5	0.671	0.868	8.4	6
6	0.695	0.837	10.6	1
7	0.787	0.681	7.8	198
8	0.529	1.237	131	203
9	0.518	1.278	293	3598
10	0.435	1.643	2519	49
11	0.449	1.552	4333	1611
12	0.457	1.478	8054	9986
13	0.455	1.506	18706	–

One must note that the applied models are not well-suited to programmes like ours where only a few errors are found. The models are better suited to very large programmes where a large number of errors are found.

CONCLUSIONS

In this section our conclusions about the various phases of the project are summarized. First, it became evident that programme specification is a weak link in the software development chain, and we think that one of the causes for this weakness was the lack of a formal specification method. Concerning the programming phase, the modular structured programming method should be used in a manner such that the variables of each module are classified and each module has well defined entry and exit conditions. This feature will also ease the programme analysis. The method of an independent inspector checking the programme correctness during all phases has the advantage that programming errors are more likely to be found at an early stage in the programme development. The use of dual programming, however, seems to be a better way of checking the complete freedom from errors in a programme. One reason for the effectiveness of this method is that it makes it possible to execute the two programmes simultaneously on a large number of test data. The introduction of artificial errors is a good way of checking the effectiveness of the programme verification methods. Statistical methods for reliability assessment of software should be used carefully, since our experience indicates that different methods may give quite diverging results. However, some confidence in the programme reliability numbers can be obtained by using several of the statistical methods and checking them for consistency.

Our general conclusion is that in any "reliable software development project" the entire effort should be viewed as one integral project. All phases of the development, from specification to verification should be based on the same underlying structure. Following a procedure of this type will produce a more reliable final product and will also significantly reduce the effort required in the programme validation phase.

REFERENCES

Mills, H.D. (1972). On the statistical validation of computer programmes. IBM paper FSC 72-6015.

Moranda, P.B. (1975). Prediction of software reliability during debugging. Proc. 1975 Annual Reliability and Maintainability Symposium, pp. 327 - 332.

Nassi, I., Schneidermann, B. (1973). Flow Charting techniques for structured programming. SIGPLAN Notices (ACM) V8, No. 8, 12 - 26.

Sukert, A.N. (1976). A software reliability modelling study. RADC-TR-76-247. Rome Air Development Centre, Griffiss Air Force Base, New York.

DISCUSSION

Trauboth: I have a question on the specification. I got the impression that these specifications were in natural language. Did you look at some other techniques like the 'CASCADE System' in Norway? Have you checked the possibility of applying this method for specifying your system?

Lathi: No. We decided to use only natural language.

Voges: Can you say something about the size of the programs in PASCAL and FORTRAN?

Lathi: The FORTRAN program had 22 000 words with data, the program code was about 11 000 words. The PASCAL program was almost twice as large. The FORTRAN program had 500 lines and the PASCAL program 1500 lines.

Quirk: Can you mention something about a comparison between FORTRAN and PASCAL?

Lathi: We cannot actually give any values since the PASCAL team didn't use PASCAL before.

A SURVEY OF METHODS FOR THE VALIDATION OF SAFETY RELATED SOFTWARE

U. Voges and J. R. Taylor*

Kernforschungszentrum Karlsruhe GmbH, Institut für Datenverarbeitung in der Technik, Postfach 3640, D-7500 Karlsruhe, Federal Republic of Germany
*Risoe National Laboratory, Electronics Department, DK-4000 Roskilde, Denmark

Abstract. This paper is written on behalf of the European Workshop on Industrial Computer Systems (Purdue Europe), Technical Committee on Safety and Security (TC 7). Its subgroup on software has studied the problems of safety related software for the last five years. This paper reports on most of this work that has been carried out by individual members of the group and that has been discussed thoroughly within this group. The main areas which are covered are specification, program construction, statistical and analytical testing methods and the use of tools. The paper closes with some conclusions and some remarks concerning future work.

Keywords. Program analysis; reliability; software diversity; software engineering; software testing; software validation; statistical testing.

INTRODUCTION

This paper is written on behalf of the European Workshop on Industrial Computer Systems, Technical Committee 7 on Safety and Security. Over a period of five years, the group has studied the problems of developing and of testing software, and of demonstrating its correctness. These are important steps if licenses are to be achieved for computer based safety systems, for example for railway traffic control, or for nuclear power plant safety systems. Several members of the group are actively engaged in developing such systems, and in the development of the theory of testing. This has provided a good opportunity to compare different approaches to testing. The strengths and weaknesses of different approaches are compared here.

The objective of this paper is to report on those topics where some own work has been done and some experience has been gained. The paper does not intend to present much formal and mathematical details, but tries to help in understanding the discussed methods and their limitations. Theoretical details are available in the numerous working papers of the Technical Committee and the referenced publications.

A separate paper deals with the problem of hardware errors (Schwier, 1979). Parts of this paper were presented in an earlier version at the IFAC Congress in Helsinki (Taylor and Voges, 1978).

SPECIFICATION

The specification is the first phase in the system development process. Therefore much emphasis has to be put into this stage, because future errors can be impeded or facilitated. Every error which is made in the specification review, will probably not be detected before the final acceptance testing. Some studies have shown that as many as 60% of all software errors can arise in specification (Boehm, McClean, and Urfig, 1975; Endres, 1975).

Therefore some work was done by TC 7 members to draw attention towards the specification process. The process of specification has to be formalised, and the use of tools which help during this phase seems to be very important. Different methods and tools for the software specification and design were inspected and evaluated (Ludewig and Streng, 1978). Some experience has been gained by applying a tool which is already used by some members (Trauboth, 1978). Other work used formal logic to specify the requirements of complex real-time systems (Quirk and Gilbert, 1977). This method can also be used for proof of correctness and analysis of the specifications.

The work in this area has not come to a final conclusion yet. Further work has to be done to provide system engineers and programmers with common tools and methods to specify the system requirements in a clear and (almost) error-free way.

PROGRAM CONSTRUCTION

Almost all of the methods to be described for software validation benefit greatly from careful program construction. In particular, the methods of path testing and program proof benefit greatly from techniques of structured programming (Ehrenberger and Taylor, 1977). For parallel or real time programs the methods described by van Horn (1968) are particularly important, since they reduce the number of program states to be examined in testing to just those which are relevant to program logical performance.

SOFTWARE DIVERSITY

Application of statistical formulae (Ehrenberger and Ploegert, 1978) reveals the fact that for larger systems, and particularly those with large data bases, the amount of testing required to validate a high reliability with a high degree of confidence is too great to be practical. An approach which allows a lower reliability target to be set is to provide diverse software (e.g. different computers, different programmers, different methods). In this way, a highly reliable system can be constructed with less reliable software.

For safety shutdown systems, comparison or voting between diverse software can be accomplished by (fail safe) hardware. For more complex systems, comparison of results can be more difficult, requiring either (partially) synchronised, redundant computers, or software cross checking of results.

The strength of software diversity is that it reduces the necessary reliability because there is a chance to detect dangerous failures in the separate programs during operation without catastrophe. This allows improvement of safety during operation.

The main remaining weakness, which prevents diverse programming being adopted as a complete solution to software reliability problems, is that like other programs, diverse programs are subject to errors arising in specification. Specification errors can give rise to common cause failure in that both programs are affected in the same way. The fact that two programming teams carry out the programming means that there is a higher probability of detecting specification errors. But members of TC 7 have observed such common specification errors in diverse programs in their own projects (Dahll and Lahti, 1979; Gmeiner and Voges, 1979).

At present there is no knowledge of the extent to which common cause failures will arise.

When compared with systematic or random statistical testing, then the relative advantages of diverse programming depend on the way in which testing is carried out. If test results are checked by a relatively weak technique such as proof reading the diverse programming would be expected to give better reliability. If a strong technique for checking test results, such as execution with a plant simulator, or execution in a full scale experiment with real plant components, then the testing approach gives a better check for specification and for programming errors. This simply because the testing must be more thorough (reduction in amount of testing necessary is an advantage of diverse programming). But the highest reliability, and greatest certainty in the results, arises when both systematic or statistical testing, and diverse programming are used.

Using diverse programs as a means of checking test results, checking one program against the other, is a convenient way of detecting programming errors, but is not so effective as execution with a plant model or with a pilot plant in detecting inadequacies in the specifications.

A subtle point in discussion of software diversity is that errors can accumulate in program data. This accumulation can lead to a gradual increase in failure rate. The failure is detected as a result of the redundancy, but the system availability is reduced. Therefore there must be periodic checking, or continuous redundant checks, to remove errors, if a high degree of availability, as well as safety, is to be achieved.

STATISTICAL METHODS

The problems of software testing are well known. Even simple programs often involve inconveniently large numbers of tests. For example one (small) safety program, for testing all combinations of input data, required 10^{28} tests (de Agostino, Putignani, and Bologna, 1977). Therefore it is important to have methods which help to reduce the necessary amount of testing.

Supervising the Debugging Process

The only methods which provide directly a reliability or risk estimate for software are statistical testing techniques. Methods for assessing probability of software failure using models of the debugging process (e.g. Jelinski and Moranda, 1973; Shooman, 1972) seem to be most useful in judging progress in debugging and predicting availability for ordinary computing systems (Hallendahl, Hedin, and Oestrand, 1975). When judging safety of highly reliable systems such models are generally too simple, compared with the actual debugging process, and make assumptions which tend to neglect second order effects such as load sensitivity, data error accumulation and misunderstanding of requirements in rare high risk situations.

They also assume that any programming error can be found during an arbitrary test run, what of course is not true in general, since some errors are only detectable after others have been removed. In addition it can be said that during debugging test data are mostly chosen according to some intuitive understanding of the test object and not following some random selection procedure. So it is not surprising that Moranda (1975) reports on a debugging process which did not fit into the model assumptions. A similar experience comes from Dahll and Lahti (1979), where several models were compared by applying them to the debugging of two diverse programs. It turned out that some could be used, others not.

Besides this the possible accuracy for estimating the related model parameters from the observed debugging results is not too high. Therefore the reliability predictions to be gained are not very accurate. Last but not least all the known models fail if no bugs are found and the most optimistic reliability parameters to be gained are insufficient for most safety applications.

Statistical Test of the Debugged Program

An alternative to such direct modelling is to use post debugging tests using randomly selected data (Ehrenberger, Hofer, and Hoermann, 1975). The results of such testing can be checked by applying the result to a simulation of the plant to be controlled, or by writing diverse programs (different programmers, programming languages, and/or computers).

The strong point of this kind of method is that it is independent of any assumptions about the way in which programs are constructed, and it provides examples of output checked against a diverse standard (another program or a plant simulation).

One problem with this approach is the enormous amount of testing required to provide reliability estimates to which a high level of confidence can be attached. Structuring the software and applying tests to parts of it is one way in which the situation can be improved (Ehrenberger, Hofer, and Hoermann, 1975). Another method is to distribute software among several microcomputers, and to test these independently, in addition to performing integrated tests (Cooper and Wright, 1977).

A more fundamental problem which has been identified is that the pattern of inputs to a program during testing may differ very greatly from the pattern applied during actual use. The theory of stratified testing has been adopted to program testing (Ehrenberger and Ploegert, 1978) and allows the tests to be modified according to the eventual load, provided that this load can be predicted.

Another encouraging aspect comes from the different frequencies of traversal of specific program parts. Some parts may be traversed very often during one program run, because they are inner parts of loops or they are subroutines, called from several places. So it can happen, that the required high number of test runs to show a certain reliability figure is not fulfilled for the whole program, but for some of its parts. In some cases it may even be possible to show that such a frequently used part has been completely tested with a sufficiently high probability. The quoted report gives the related mathematical formulae.

General Evaluation

Any statistical testing technique depends on certain assumptions, such as assumption of uniformity of error distribution within a program, or lack of coupling between the error cause mechanism and the way in which the program is used or on a certain distribution of test data. During testing it may be difficult or even impossible to provide the appropriate distribution of test data, since some necessary parameters are not known with the required accuracy. This can enforce a higher number of test runs.

An important step in the development of statistical testing methods is to 'test the methods themselves'. By applying different methods to the same set of errors, it should be possible to determine the importance of these assumptions.

ANALYTICAL METHODS

Various analytical software validation methods decompose the investigated software into small subunits. The purpose for this decomposition is manifold, e.g.
- the structure of the program is visualized,
- in some cases the more intensive validation can be made easier for the subunits,
- the minimally needed number of test cases is reduced very much.

Depending on the used approach the emphasis lies on consideration of the data flow or of the control flow. Such independencies of subunits can normally be shown by purely formal investigation of the program code, without any intuitive understanding of the purpose of the individual software parts.

Path Testing

The theory and techniques developed in program proving can be adapted to make various kinds of systematic testing possible. As an example sets of test data which ensure that all arcs (sections of program between branch points) in a program or all paths through a program can be developed (Seifert, 1976; Taylor and Bologna, 1977).

A typical method of deriving such sets of test data is to place an assertion at the end of a program, as in proof of correctness, but in this case, the assertion is 'neutral', consisting of just 'the program terminates here'. Then program statement axioms can be used to derive assertions or predicates which must be true at earlier points in the program, in order for correct termination to be possible. By keeping the predicates associated with each path through the program separate, sets of simultaneous equations or inequalities can be derived, which have to be solved to provide sets of test data covering each possible path through a program.

The main difficulties with this method are the large numbers of possible program paths, and the difficulties of simplifying and 'solving' the path predicates to give sets of test data. Both of these problems have shown to be simpler, for the kind of safety programs studied, than for programs in general. A major advance has been minimisation of the number of paths which need to be tested, based on data flow considerations (Taylor and Bologna, 1977).

Is the method of path testing, using systematically derived test data, a complete solution to the software reliability problem? We have discovered three kinds of errors which are not detected by path testing. First, the currently used axioms for programming languages do not take into account problems of arithmetic overflow and arithmetic accuracy in general. To do so does not appear difficult, but would require treating arithmetic expressions in decomposed form, with arithmetic operators treated in the same sequence as they would be executed in compiled code. To understand the second problem, one must consider that with each path through a program a certain domain of input data is associated (for example, all executions with prime numbers as input follow one path, all those with none prime number input follow another). A programming or specification error arising because some domain distinction (that is, a set of special subcases of some part of a path domain) has been forgotten will not necessarily be detected by this technique. For example, a safety program might be constructed to sound an alarm if either temperature or pressure exceeded a certain limit. It might be forgotten that still lower limits were necessary if both pressure and temperature rose at the same time. A program path for this contingency might not exist, and testing might then only inspect part of the input domain where pressure and temperature changed independently.

The problem can be reduced by testing programs for all points along path domain boundaries, in which case only errors involving omission of more than one branching statement can be overlooked. Errors in control flow decisions, such as those arising from substituting '<' for '<=' in an IF statement, can be detected by this method.

There is some difficulty with the path testing approach to validation in general, however, because of the third problem on our list. A test may be applied within a path domain yielding a result which is correct, even though the computation performed along that path is in general erroneous. The correct result is a 'coincidence' arising from a special choice of the path testing data. The problem for statistical testing within a domain is a problem of how 'rare' such coincidences might be. This does not appear to be a problem for the safety relevant programs so far tested, provided that both statistical testing and path testing are used (correctness has been checked for several examples using proving techniques). However, the problem indicates the need for statistical testing.

Symbolic Evaluation

An additional method which can be useful is the symbolic evaluation. The program is executed not with actual data but with symbolic input values. Each arithmetic computation and each assignment is done symbolically, and the different paths of the program are associated with different outcomes of the conditional statements. The final result is an expression in which the output values are a function of the input values and additional constants, arithmetic and logic operators and relations. This function can be used for comparison with the specifications and represents a general description of the mapping of the program.

Depending on the size of the program and the complexity of computations and dependencies this function may be hard to understand and to interpret. But e.g. errors of kind three in the preceding paragraph can be detected by symbolic execution.

The method of symbolic evaluation has some points in common with the above mentioned method of path predicate calculation.

Manual Program Analysis

The different methods for analysing and evaluating programs by automatic tools are not completely satisfactory up to now. They give some confidence in performed test, but they lack completeness, or they do at least not provide the knowledge that the performed tests were complete. So some additional work by hand is required. Certainly also the total analysis can be performed manually, as e.g. described by Ehrenberger, Rauch, and Okroy (1976) and by Ehrenberger and Pur-Westerheide (1979). This however is not recommended because it takes much time and it is very error prone.

Use of Tools in the Process of Validation - An Example

Some of the methods discussed in this paper are supported by tools like RXVP (Miller, Paige, and Benson, 1974), FACES (Ramamoorthy and Ho, 1974), or SADAT (Amschler, 1976) which can be used to provide the user with some of the desired results more or less automatically. The extent of information given by the automated tools varies between the different systems. As an example we take the system SADAT which consists of the following main parts:

- static analyzer
- instrumentation
- test case selection
- path predicate generator
- execution verifier.

The static analyzer as the basic step sets up the data base for the following steps and does some fundamental checking (e.g. initialisation of variables, dead code, type consistency). The code can be instrumented in order to receive a histogram of the execution of the program. The test case selector selects out of the set of possible paths a nearly minimal subset which guarantees that every decision is activated and every statement is executed at least once. After that the appropriate path predicates are computed, resulting in conditions for the selection of test data.

This procedure shall be illustrated with a very simple program (Fig. 1).

```
1    integer a,b
2    real z
3    read a,b
4    if (a .eq. 0 .or. b .eq. 0)
5       then z=0
6       else
7          if (a .lt. b)
8             then z=a/b
9             else z=b/a
10         endif
11         if (z .lt. 0)
12            then z=-z
13         endif
14   endif
15   write z
```

Fig. 1. Example program

Following paths through the program are possible:

(1) 1 2 3 4 5 14 15
(2) 1 2 3 4 6 7 8 10 11 12 13 14 15
(3) 1 2 3 4 6 7 9 10 11 13 14 15
(4) 1 2 3 4 6 7 8 10 11 13 14 15
(5) 1 2 3 4 6 7 9 10 11 12 13 14 15

The respective path predicates are:

(1) (a .eq. 0 .or. b .eq. 0)

(2) .not. (a .eq. 0 .or. b .eq. 0)
    .and. (a .lt. b)
    .and. (a/b .lt. 0)

(3) .not. (a .eq. 0 .or. b .eq. 0)
    .and. .not. (a .lt. b)
    .and. .not. (b/a .lt. 0)

(4) .not. (a .eq. 0 .or. b .eq. 0)
    .and. (a .lt. b)
    .and. .not. (a/b .lt. 0)

(5) .not. (a .eq. 0 .or. b .eq. 0)
    .and. .not. (a .lt. b)
    .and. (b/a .lt. 0)

Paths 1-3 give a coverage of the program, every segment and each decision of the program is activated at least once. The related test data could be:

(1) a=0, b=7
(2) a=-2, b=14
(3) a=13, b=4

The complete set of all paths is achieved with the paths 1-5. The additional test data could be:

(4) a=5, b=27
(5) a=17, b=-3

If the program is exercised with these test data, the following histogram will be shown:

program line number	executed in test number
1	1 2 3 4 5
2	1 2 3 4 5
3	1 2 3 4 5
4	1 2 3 4 5
5	1
6	2 3 4 5
7	2 3 4 5
8	2 4
9	3 5
10	2 3 4 5
11	2 3 4 5
12	2 5
13	2 3 4 5
14	1 2 3 4 5
15	1 2 3 4 5

PROOF TECHNIQUES

Proof of Satisfaction of Specifications

Methods for proof of correctness, or rather for proof of satisfaction of formal specifications, have been investigated by some members of the group (e.g. Taylor, 1974). The methods of proving are not considered to be of much help alone in guaranteeing software safety, because the step of deriving formal specifications is one of the most error prone of all programming stages (Goodenough and Gerhart, 1975). The process of proof proceeds according to fig. 3 with a stage of formalisation from an intuitive, unexpressed specification, rather than by the simple process of fig. 2.

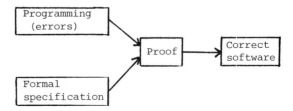

Fig. 2. Ideal Process of Program Proving

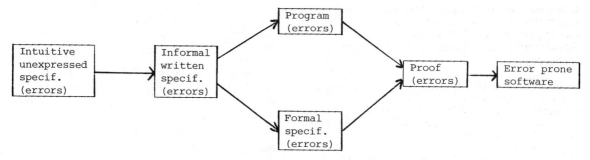

Fig. 3. Normal Process of Program Proving

Proof techniques remain the primary means at present, by which errors can be detected in parallel programs. In this case the number of cases to be tested either statistically or systematically grows very rapidly. An important point is that the techniques of proof benefit greatly from the general simplicity of safety related systems (Taylor and Bologna, 1977).

Proof of Loops - An Example

Program proof techniques, especially the proofs by induction, are possibly the only means of showing the correctness of general WHILE-loops. We want to demonstrate this with a short example. The program for integer division by successive substraction shall serve as an illustration:

```
1 read x,y
2 r=x
3 q=0
4 while (y<=r)
5     r=r-y
6     q=1+q
7 endwhile
8 write q,r
```

The input values x and y are both integers and greater zero. It is to be proven that after program execution we have the output values r and q with:

$r<y$ and $x=r+y*q$

a) If $x<y$, we do not execute the loop and $r<y$ and $x=r+y*q$ is true with the initialisation assignments r=x and q=0.

b) If $x=y$, the loop is executed once. The following assignments take place: r=x; q=0; r=r-y; q=1+q. Therefore r=0 and q=1, which shows that $r<y$ and $x=r+y*q$ holds.

c) If $x>y$ we have the initialisation
$r_0=x$
$q_0=0$
and the first loop traversal
$r_1=x$
$q_1=1+q_0$
With this $x=r_1+y*q_1$ is true, since $x=x-y+y*1$.

With the assumption that
$x=r_n+y*q_n$ and $r_n>=y$
is true after the n-th loop traversal, then
$x=r_{n+1}+y*q_{n+1}$
is true after the (n+1)-th loop traversal, since
$x=r_n-y+y*(1+q_n)$ $(r_{n+1}=r_n-y; q_{n+1}=1+q_n)$
$=r_n+y*q_n$
which holds according to the assumption.

The loop ends since r is a finite number at the beginning and is decreased with each loop traversal and y is positive.

The example shows that program proofs are not simple. As a rule the effort for proving is higher than the effort for programming. An essential difficulty is to derive the loop assertion or loop invariant from the code. This cannot be done automatically and requires full intuitive understanding of the related piece of code (Gries, 1976). Program proofs can contain errors. In addition to this the program can fulfil the assumed assertion but have additional properties, which are not covered by the assertion.

SUMMARY AND FUTURE DEVELOPMENTS

The present outlook for safety related software validation is bright. The way seems clear for a combination of statistical and path testing e.g. statistically selected data could be applied to search for forgotten cases. The resulting statistical theory is simpler than for program errors in general, on the assumption, that each test is unequivocal. A combination of path testing and statistical testing appears to cover many of the problems which can be raised in objection to any one method. There remains a problem of certain special kinds of program, which along some path have an erroneous computation, but which give the correct result for most data. Experimental work is being undertaken
- to check to what extent the coincidence problem is significant for safety programs,
- to validate the theory of statistical testing.

For multiprogrammed and general multiprocessor safety systems, many problems remain. Proof techniques (in combination with special software construction methods) are the only ones which provide a solution to problems of relative timing, and for these, problems of achieving an accurate and complete formal specification are large. In any case, such methods will need to be supported by testing techniques.

Given that theoretical results and experiments in testing inspire a high level of confidence, but acknowledging that there remain some theoretical problems, the techniques of diverse programming provide us with an extra level of insurance (Chen and Avizienis, 1978). If a serious reliability problem does arise, because of a theoretical oversight, then we should obtain warning of it within one diverse program, before it becomes a safety problem.

The major problems with safety related software then seem to be:

- potential for errors in original specifications,
- potential for common mode errors in diverse programming.

Both of these problems focus attention on the way programming errors arise. They do not seem to be very much different in nature from existing design errors in hardware safety systems.

Finally, in fig. 4 the correlation between different errors and their related detection methods is summarized.

ERROR TYPE	DETECTION AND/OR PREVENTION METHODS
Software errors in specially constructed programs	
Errors in a path data transformation	Automatic path testing, if transformation is a polynomial; hand assisted path testing otherwise.
Errors in control statements	Automatic path boundary testing.
Omission of special cases	Derivation of tests from specifications.
For more general programs	
Software errors in general	Manual path testing, statistical testing.
Parallel programs, real time programs	Proof, where possible systematic execution sequence tests. Program timing analysis.
All software errors	Software diversity provides backup.
Specification errors	Simulation, design review, specification analysis.

Fig. 4. Relation Between Errors and Related Detection Methods

ACKNOWLEDGEMENT

The authors are grateful to their fellow members of TC 7, especially to the members of the software subgroup. Special thanks go to S. Bologna, M. J. Cooper, G. Dahll and W. Ehrenberger for their written contributions.

REFERENCES

de Agostino, E., M. G. Putignani, and S. Bologna (1977). Aspects of the Software Design for the Computerised Protection System for the Fast Reactor Tapiro. IAEA/NPPCI Spec. Meeting on Software Reliability for Safety Systems in Nuclear Power Plants. Pittsburgh.

Amschler, A. (1976). Test Data Generation in an Automated Testsystem for FORTRAN-Programs (in German). Diplomarbeit, Universität Karlsruhe.

Boehm, B. W., R. L. McClean, and D. B. Urfig (1975). Some Experiments with Automated Aids to the Design of Large-Scale Reliable Software. IEEE Trans. Software Eng., 1, 125-133.

Chen, L., and A. Avizienis (1978). N-Version Programming: A Fault Tolerance Approach to Reliability of Software Operation. FTCS-8 Toulouse.

Cooper, M. J., and S. B. Wright (1977). Some Reliability Aspects of the Design of Computer Based Reactor Safety Systems. IAEA/NPPCI Spec. Meeting on Software Reliability for Safety Systems in Nuclear Power Plants. Pittsburgh.

Dahll, G., and J. Lahti (1979). An Investigation of Methods for Production and Verification of Highly Reliable Software. IFAC-Workshop SAFECOMP '79, Stuttgart.

Ehrenberger, W., E. Hofer, and H. Hoermann (1975). Probability Considerations Concerning the Test of the Correct Performance of a Process Computer by Means of a Second Computer. IFAC Congress, Boston.

Ehrenberger, W., and K. Ploegert (1978). Einsatz statistischer Methoden zur Gewinnung von Zuverlässigkeitskenngrößen von Prozeßrechnerprogrammen. KfK/PDV 151.

Ehrenberger, W., and P. Pur-Westerheide (1979). Analytical Software Verification. Symposium on Avionics Reliability, its Techniques and Related Disciplines. Ankara.

Ehrenberger, W., R. Rauch, and K. Okroy (1976). Programanalysis - A Method for the Verification of Software for the Control of a Nuclear Reactor. 2nd Internat. Conf. on Software Eng., San Francisco.

Ehrenberger, W., and J. R. Taylor (1977). Recommendations for the Design and Construction of Safety Related User Programs. Regelungstechnik, 25, 2, 46-53.

Endres, A. B. (1975). An Analysis of Errors and Their Causes in System Programs. IEEE Trans. Software Eng., 1, 140-149.

Gmeiner, L., and U. Voges (1979). Software Diversity in Reactor Protection Systems: An Experiment. IFAC-Workshop SAFECOMP'79. Stuttgart.

Goodenough, S. L. Gerhart (1975). Toward a Theory of Test Data Selection. Internat. Conf. on Reliable Software. Los Angeles.

Gries, D. (1976). An Illustration of Current Ideas on the Derivation of Correctness Proofs and Correct Programs. 2nd Internat. Conf. on Software Eng. San Francisco.

Hallendahl, G., A. Hedin, and A Oestrand (1975). A Model for Software Reliability Prediction for Control Computers. TRITA-TTS-7502. Department of Telecommunications, Royal Institute of Technology. Stockholm, Sweden.

van Horn, E. C. (1968). Three Criteria for Designing Computing Systems to Facilitate Debugging. Commun. ACM, 11, 360-365.

Jelinski, Z., and P. B. Moranda (1973). Applications of a Probability Based Model to a Code Reading Experiment. IEEE Symposium on Software Reliability.

Ludewig, J., and W. Streng (1978). Methods and Tools for Software Specification and Design - A Survey. Purdue Europe TC 7. Working Paper No. 149.

Miller, E. F., M. R. Paige, and J. P. Benson (1974). Structural Techniques of Program Validation. COMPCON 1974, pp. 161-164

Moranda, P. B. (1975). Prediction of Software Reliability During Debugging MDA. MDAC paper WD 2471.

Quirk, W. J., and R. Gilbert (1977). The formal specification on the requirements of complex real-time systems. AERE-R 8602.

Ramamoorthy, C. V., and S. F. Ho (1974). Fortran Automated Code Evaluation System (FACES), Part 1. ERL-M 466. Univ. of California.

Schwier, W. (1979). Overview of Hardware-Related Safety Problems of Computer Control Systems. IFAC-Workshop SAFECOMP 1979. Stuttgart.

Seifert, M. (1976). SADAT, Ein System zur automatischen Durchführung und Auswertung von Tests. KfK-Ext. 13/75-5.

Shooman, M. L. (1972). Probabilistic Models for Software Reliability Prediction. Internat. Symposium on Fault-Tolerant Computing. Newton.

Taylor, J. R., and S. Bologna (1977). Validation of Safety Related Software. IAEA/NPPCI Meeting on Software Reliability for Safety Systems in Nuclear Power Plants. Pittsburgh.

Taylor, J. R. (1974). Proving Correctness for a Real Time Operating System. IFAC-IFIP Workshop on Real Time Programming. Budapest.

Taylor, J. R., and U. Voges (1978). Use of Complementary Methods to Validate Safety Related Software Systems. 7. Triennial World Congress of the Internat. Federation of Automatic Control. Helsinki.

Trauboth, H. (1978). Introduction to PSL/PSA. Purdue Europe TC 7 Working Paper No. 151.

DISCUSSION:

<u>Parr:</u> How sure can one be that in a very large program (100k or 200k words of program) no more error is left?

<u>Voges:</u> If you have very large programs then you must have additional features like fault-tolerant software or error-detection-mechanisms included in the software in order to cope with remaining errors.

<u>Keutner:</u> You mentioned three tools for automatic testing: FACES, SADAT an TEVERE. Can you say something if they are restricted to specific languages or systems and where they are used?

<u>Voges:</u> RXVP, FACES an SADAT are used FORTRAN-programs. RXVP is written in FORTRAN and is commercially available; a derivate of FACES can only be used in a service center in the USA. SADAT is developed in our institute, is written in PL/1 and is available due to negotiations. TEVERE is written in LISP and is available from Mr. Taylor in Riso or from Mr. Bologna in Rome.

AN EXPERIENCE IN DESIGN AND VALIDATION OF SOFTWARE FOR A REACTOR PROTECTION SYSTEM

S. Bologna, E. de Agostino, A. Mattucci, P. Monaci and M. G. Putignani

Lab. Automazione e Controllo Impianti CNEN, Casaccia, Rome, Italy

Abstract. The paper describes the full life cycle process of design, development and validation, which has been followed for the production of software for an experimental computerized reactor protection system at Casaccia Center. The aim of the paper is to put in emphasis phase-by-phase the criteria which have been followed, the verification procedures adopted and the errors discovered. Additionally the system testing activity is described.

Keywords. Computer control; reliable software; software validation; program testing; system testing.

INTRODUCTION

The paper describes the full experience gained in the design, implementation and validation of the computerized protection system for the TAPIRO fast reactor at Casaccia-Center - Rome. The aim of the work was the investigation of methods to ensure the production of reliable software in terms of design qualification and acceptance testing. The design basis for the System is described in De Agostino, Mattucci (1977); the protection algorithm and software design are reported in Monaci, Putignani (1977), whereas the implementation methodology and proving strategy are described in Bologna and others (1978).

The aim of the paper is to put in emphasis phase-by-phase the criteria which have been followed, also from organizational point of view, to improve the quality of the software.

The verification procedures adopted and the errors discovered are reported, also the system testing strategy is presented. The relationship between the different phases of the development and validation cycle is shown in Fig. 1.

SOFTWARE REQUIREMENTS

The only document available at the beginning of the project was the system requirements. This document deals both with functional requirements and performance requirements. This document is constituted of a part written in Italian language, plus a set of tables that show in the form of decision tables the relations, between the values that can be assumed from the variables and the corresponding actions.

The software design phase started directly from this document, without producing a specific software functional requirements document devoted exclusively to software developers. The lack of this phase, that normally involves people with different backgrounds, and should lead to a formal document which should provide a common medium for communications between non software development specialists and the software developers, was cause of problems during the subsequent phases in the software life cycle.

One of the major problems was due to the ambiguity and incompleteness of system requirements used as software requirements. This has implied that the people involved in the software design have been obliged to apply several times to the people that had prepared the system requirements document, or, even worse, to make some personal assumptions for covering the missed information. As consequence the system requirements has been improved or changed several times during the software life cycle, with obvious consequence on the subsequent phases. Essentially we can group all the changes apported in the requirements into three groups:

- Requirements changes due to a better knowledge of the plant: Incompleteness.

- Requirements improvement due to a bad and confused presentation of the information in the previous editions: Ambiguity.

- Requirements changes due to some doubts that came up during the code verification phase. One of the old system requirements did not agree with the safety report of the plant.

Besides the system requirement specifications, also a software validation plan has been used as starting document in the software life cycle. The plan contains:

- a detailed description of the methodology to be followed during the software design phase;
- a detailed description of the code verification activities at module level;
- a detailed description of the integrated system test with the simulation of the environment of the system.

SOFTWARE DESIGN

The first ten months of the project were devoted to creating a top down detailed system software design at module level. The design activity has been conducted in two steps: preliminary design and detailed design. For each step a corresponding design review has been carried out as shown in Fig. 1.

Preliminary design

The preliminary design has been the first step in decomposing the problem into a machine solvable form. In this phase the function of the software has been defined in terms of a set of primitive functional components. The arrangement of these components defines the processing order of the functions and the data applied to each functional component. The language adopted in this phase has been a standard flowchart language integrated with written comments. Its major weakness is the lack of formal syntax and semantic description; this has allowed the people involved in the software design activity to produce documents with very different levels of detail and conciseness. The preliminary design has proceeded to a level where the major decisions concerning the determination of the functional components of the software system, allocation and use of computing resources, interfaces with computing environment were delineated.

Preliminary design review

At the conclusion of the preliminary design, a preliminary design review has been conducted from different people that had not partecipated to the design activities. There was no formal methodology established for the review. The aim of this review was to guarantee that all the functional requirements had been fully mapped into the preliminary design according to the established methodology. For this reason the design description has been reviewed for completeness and consistency. During the review eight errors due to misinterpretation of the system functional requirements have been pointed out.

Detailed Design

In the detailed design phase, starting from the preliminary design description, the refinement process has been continued in a top-down manner until to sufficient detail to allow the coding activity. The result of this activity has been the detailed description of the interrelation of all modules that constitute the program and a detailed description of the functions executed for each module. Also in this phase, a standard flowchart technique integrated with written comment has been used, for describing both the modules interrelation and the data manipolation with control flow within the module. The document produced has been the reference document for the coding activity.

Detailed Design Review

As for the preliminary design review the detailed design review has been conducted from different people, and there was no methodology established for the review. The aim of this review work was to verify the detailed design description for completeness, unambiguity and consistency (both internal consistency and consistency with the description of the prior phase result).During the review seven errors have been discovered. These errors can be considered as design inaccuracies (e.g. variables which are not initialized, etc.)

SOFTWARE CODING

The protection program has been written in IFTRAN/3 (an extension to FORTRAN) and has been translated to RTL/2 language, a language that permits the input-output management on the ULP/12 end computer.

For this reason the program written in IFTRAN/

3 contains comment statements whenever input-output statements are necessary.

Both the protection software design description and the input-output RTL/2 procedures description are the only references for the programmer, that is program specification. At the end of coding activity, three documents have been produced:

- the first one contains all purely combinatorial modules, where for each module both module specifications and the associated program code (IFTRAN/3 containing comment statements) are presented;

- the second one contains the nuclear variables module, where both module specifications and the associated program code (IFTRAN/3 containing comment statements) are presented;

- the third document contains the whole integrated program code in ULP/12 Assembler, as result of the translation IFTRAN/3 vs. RTL/2 and the compilation by RTL/2 Compiler.

Each one of the three documents mentioned before has been used from one of the three different code verification teams; it was the basic document together with the specific software validation plan.

CODE VERIFICATION FOR THE PURELY "COMBINATORIAL MODULES"

Procedure description

In the verification procedure adopted for the purely combinatorial modules, information gained from testing the program using test data derived automatically through an analysis of the program code, have been used as an aid in proving program correctness. One can always, in fact, say that a program is correct by using the program code for deriving its own specifications. If the so derived program specifications match the original program specifications itself we are allowed to say that the program code corresponds to its specifications. For applying this procedure, program specifications must obviously be expressed in a formal way. The adopted procedure for each purely combinatorial modules of TAPIRO reactor protection program is reported in Fig. 2.

Looking at Fig. 2 we may see that the verification activity can be regarded as two activities in parallel; one is concerned with the analysis of program code looking only at its internal structure, the other is concerned with the formalization of the program specifications starting from the system functional requirements. In accord with the diversity concept the two parallel activities have been realized from two different sub-groups of the verification team in the software life cycle.

Obtained results

The above described procedure has been implemented on an host computer PDP11/34 with 64 K-words of core memory and an RSX11M operating system. The used software tools have been the system TEVERE-1 (Bologna, Taylor, 1978) for deriving path predicates, the system CERLTAB (Hogger, 1978) for decision table manipulation, and the program DIFTAB for making comparisons between tables. During the verification activity two errors in two different modules were found out. An analysis of the errors has showed that one can be classified as codification error, and the second one as program specification error, that is the module specification was incorrect. They both didn't touch the design of software system. The team did not correct errors it found. Very interesting has been on analysis of the gained results from the point of view of time and memory consuming. We have seen that for modules with less than ten control paths the all process takes a few hours and the time necessary for analysing the program code was nearly the same as the time necessary for deriving decision table and making comparisons. For modules with more than ten control paths the time necessary for analysing the program code increases much more quickly than the necessary time for covering all the other steps in the procedure. The necessary time for analysing the program code is a function of the number of possible control paths we have in the program and not simply of the number of statements. Our procedure becomes very time and memory consuming for modules with a few thousand program paths. For a module with a number of paths of the order of one thousand the code analysis for deriving path predicates takes a few hundred hours of computer time. For this reason at the moment we are still working on one purely combinatorial module with more than 10^4 possible control paths.

In Fig. 3 the most significant results obtained in the different steps of the procedure showed in Fig. 2, are reported for a very small module.

CODE VERIFICATION FOR "TIME DEPENDING MODULE"

Procedure description

A different verification procedure has been adopted for the module where nuclear variables are processed. The nuclear variables module is referred to the neutron flux measuring variables (two for low flux plus two for high flux) and to those variables which have a data flow dependency with them (in total, the module treats eleven variables).

This module has not a purely combinatorial switching structure, it has time depending properties. This is due to the nature of the phenomena which have to be treated and to the control criteria of those phenomena.

On the other hand, due to the very large number of possible paths ($\sim 10^{12}$), it was impossible to use previous automatic tools to have input cases.

The validation team has used the TASP system, developed by General Research (G.R.C., 1977). It was installed on IBM/370 computer which is connected with PDP11/34 in-house computer through a 2780 Emulator. This system is an automated program testing aid that provides in the Test Probe option the identification of program structures form source code, automatically insert software probes to record execution of a logical segment of code, collects run time data from the instrumented code, and generates post test analysis and reporting of testing coverage. Besides it identifies source statements where parameters are set, automatically inserts software probes to record values assumed during execution, collects run time data and performs post test analysis and reporting of statistics on the values as well as execution frequency of the statements.

The operational criteria of validation procedure established that both a systematic test and a functional test must be performed. The systematic test consisted on exciting all the arcs (DD-path) of the program at least once: arc-testing.

The functional test consisted of running the program with input data set simulating particular plant situations defined "interesting". During these runs a set of output variables values was created.

This output set was compared with expected values obtained from the system requirements.

The input cases have been found, step by step, by a careful study of flow-chart and considering the reports given by TASP.

Finally with this procedure all the DD-paths were executed.

In the functional test we have simulated reactor start-up, reactor shut-down and full power operation.

The following situations have been checked:

- high neutron flux in each measure range;
- high neutron flux rate;
- limit conditions of the nuclear instrumentation.

Obtained results

The arc-testing has revealed three errors: two coding errors and one program specification error.

The functional test has revealed two coding errors and one program specification error.

SYSTEM TESTING

Aim

The real time verification of the performance of the computerized protection system is necessary to demonstrate that the system, hardware and software, is able to satisfy the system requirements and therefore can be accepted by the Licensing Authorities. The system must be tested exciting its input terminals with well-known signal configurations and checking that the output results agree with the system requirements. To this end it is important to remind that the safety software is made up by three blocks executed sequentially:

- Test Program;
- Protection Program;
- Control Program.

The Test Program guarantees the integrity of the acquisition hardware, it is made up by modules, each of them realizes a different test on the interface; when a fault is detected the program execution is stopped and the watch-dog excitation is prevented. The real time performance of this program can be evidenced generating the fault conditions that stop the program and observing that:

– effectively the program stops;

– the Program Counter has the right value.

It is not important to test more than one different fault condition at a time as only the first one would be detected. Therefore the number of tests to be executed is equal to the sum of the possible fault conditions which can be detected in each of the modules. The fault conditions are obtained solving the path-predicates given by TEVERE-1 analyzer.

The Protection Program and the Control Program can be tested realizing a plant model able to simulate all the input signals of the protection system. Such model will be strongly dependent on the protection algorithm. In fact, if the algorithm is combinatorial, the plant model can be static, otherwise it must be dynamic.

In our case the protection algorithm is decomposable in functions independent each other. One of these functions, the one which implements the nuclear variables, is time dependent; therefore we have adopted a strategy for which the time performance verification of the protection system is divided in two steps.

In the first step the combinatorial portion is tested by means of a static model of the plant, keeping constant all the nuclear variables; in the second one the nuclear module is tested by means of a dynamic model, keeping constant all the other variables.

In the first step we can take advantage of the following features easily achievable only using a static model:

– testing of the protection system based on the program structure;

– use of the results obtained in the software verification phase for the results analysis;

– possibility to automatize completely the system verification.

The nuclear module must be tested by means of a dynamic model, able to simulate dangerous transients on the plant, operational procedures of the reactor start-up, sensors failures, etc. As it is impossible to analyze this module through the actual automatic tools, due to the large number of program paths, it is impossible to test it exhaustively; besides also the results analysis is to be executed manually.

In the system testing the following results have to be achieved:

1) a posteriori verification of the protection algorithm implementation through independent modules;

2) verification of time independency for the modules which implement combinatorial functions;

3) verification that the time interval fixed in the specifications for the execution of the protection algorithm is sufficient;

4) verification that the software has been compiled correctly;

5) verification that the input-output routines (in Assembler language) have the expected behaviour and are addressed correctly;

6) verification that the hardware meets the requirements for the protection system.

Environment and Procedures of Testing

As it has been said before, the system testing has to be done inserting the protection system in an environment which simulates the nuclear plant.

This environment is easy to be realized using a hybrid computer like EAI Pacer 700 computer available at Casaccia Center; it is composed with a digital computer EAI Pacer 100 (word length 16 bits, core store 32k) completely interfaced with the EAI 781 analog unit. Therefore it is possible to generate the analog and logical signals for the protection system under control of Pacer 100 digital unit. The system testing is done on an only channel of the protection system, as software and hardware of the two channels are equals.

The description of the hardware used is shown in Fig. 4.

The analog signals are generated in the analog computer, while the logical ones are obtained in a particular card (PLAYCARD) with connects directly the EAI Pacer 100 with the protection channel; the Playcard makes possible also exchanges of information between the computers.

The system testing cycle conceptually may be divided in four steps:

a) production of the input test cases;

b) execution of system runs;

c) results analysis;

d) documentation.

The step a) is obtained compacting, by means of the in-house computer PDP-11, the test cases corresponding to each module and the respective results. In this way an arc testing is realized and the total number of tests to be executed is equal to the paths number in the largest module. The test cases and the respective foreseen results are recorded on a backup (card), directly readable by the hybrid computer. During the step b) computer EAI Pacer 100 reads a test case, sets and checks the analog and logical values and allows the execution of the protective algorithm through the ULP-12 Flag.

At the end of each ULP-12 elaboration the obtained results, through the Playcard, are sent to EAI Pacer 100; when the scheduled number of protective algorithm executions is reached, Pacer 100 stops ULP-12 and starts the step c).

In this step the values received from ULP-12 (safety actions and alarms) in the various executions are compared among them and to the foreseen values; besides Pacer 100 checks that the values have been sent at time intervals of 100 msec.

The step d) is executed automatically by Pacer 100 on a line printer. The test case analyzed, the results obtained and a judgement on the outcome of the test case are recorded.

If the judgement is negative, an analysis is to be performed to find the reasons of the encountered diversities. This analysis is made easy by the use of the automatic tools described before.

In Fig. 4 and 5 are shown the steps of the system testing in the hybrid computer and in the protective channel.

REFERENCES

Bologna, S., E. De Agostino, D. Manzo, A. Mattucci, P. Monaci, M.G. Putignani, C. Santucci, N.M. Stephanos (1978). Implementation and Validation of a Computerized Reactor Protection System. EHPGM on Applications of Process Computers in Reactor Operation, Loen, 4-9 giugno.

Bologna, S., J.R. Taylor (1978). A Software system for programs Testing and Verification. Riso - M-1992.

De Agostino, E., A. Mattucci (1977) Design of a computerized protection system for the fast reactor TAPIRO. EHPGM on Process Supervision and Control in Nuclear Power Plants, Fredrikstad 6-10 June.

G.R.C. (1977) TASP User's Manual - General Research Corporation - December.

Hogger, E.I. (1978). CERLTAB - A Software Aid to Correct Logic Design and Programming Using Decision Table. Central Electricity Research Laboratories RD/L/N43/78.

Monaci, P., M.G. Putignani (1977). Design of reliable software for a computerized protection system. EHPGM on Process Supervision and Control in Nuclear Power Plants, Fredrikstad 6-10 June.

DISCUSSION

Mullery: If you found an error, what did you do in retesting.

Bologna: When an error was found in a module the verification team did not correct this error. The module was given back to the team leader. The design team changed the code. Then the module was given to the verification team and the verification started again.

Parr: Who wrote the validation plan? At what time and how?

Bologna: The validation plan was written by the validation team at the beginning of the project.

Levene: Can you say a few words about the form of the reviews at the end of each phase?

Bologna: There was no formal methodology for these reviews. The aim of the reviews was to look at the consistency of the documents in relation to the documents of the previous phases. In the preliminary design we had to look if the system requirements were mapped. In the detailed design the aim of the review was, to look if the preliminary design concept was mapped. We have used no formal language in the design phase and for the requirements-specification. So it was not possible to use automatic tools in this phase.

Maki: How many people were involved in the project for how many years? How many of them have experience in reactor engineering?

Bologna: The project took 2,5 years. We had the team leader and two people working on the design of the software system. Three people were working on verification at module level, two people are now working on the system test. In the meantime we have not only developed the software for the protectional system (2k IFTRAN-lines) but also the system of TEVERE-1 in cooperation with Mr. Taylor from Risö National Laboratory. We also spent some time to get familar with the other tools. In the teams there has been one reactor-engineer, the others are software engineers.

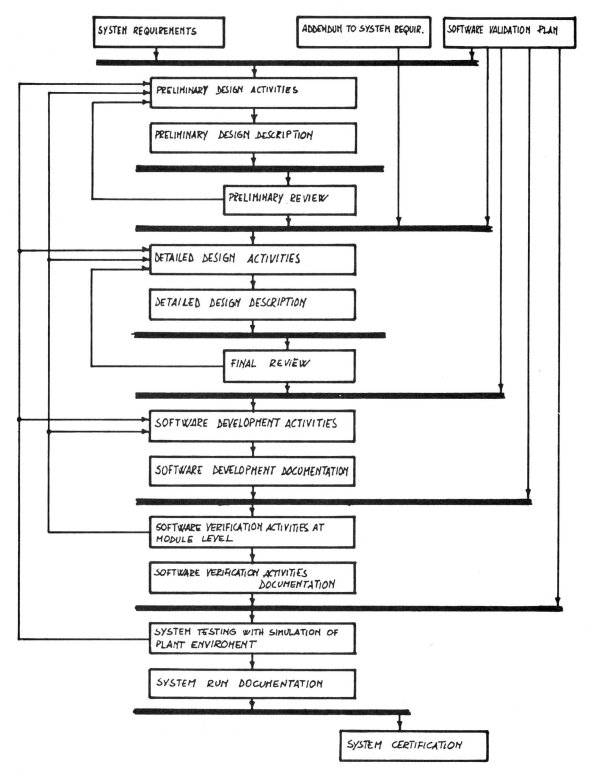

Fig.1 Design, Development and Validation Process Followed in Developing Tapiro Protection Program

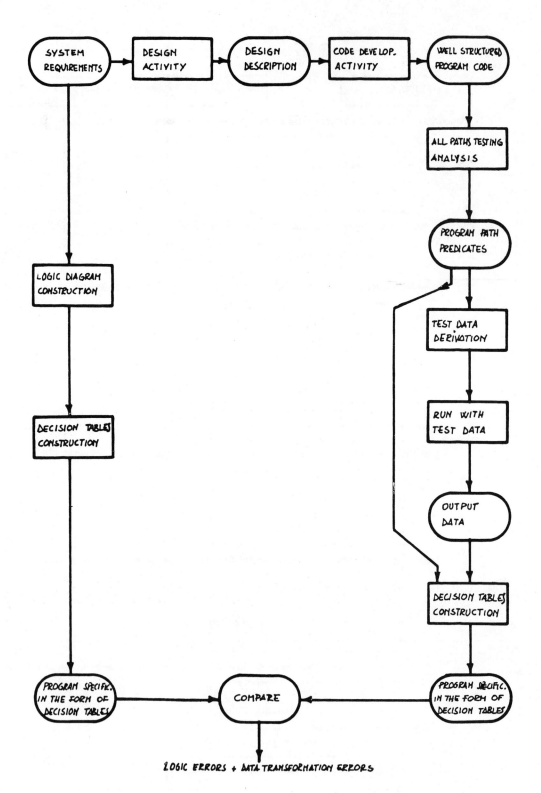

Fig.2 Code Verification Procedure for
Purely Combinatorial Modules

```
COMNT % MOD71:KV96(L,4A)-KV17(L,2A) %;
COMNT READL (KV96,96);
      IF (KV96 .EQ. 1)
         I9617=1
      ELSE
         I9617=0
      END IF
COMNT READL (KV17,17);
      IF (KV17 .EQ. 1)
         IF (I9617 .EQ. 1)
            MSK17=0
            JST17=0
         ELSE
            MSK17="4000
            JST17=1
         END IF
      ELSE
         MSK17=0
         JST17=0
      END IF
      END
END
```

FIG. 3a PROGRAM CODE

```
PROGRAM
(; (IF (EQ KV96 1 )(: I9617 1 )(: I9617 0 ))(IF (EQ KV17 1
 )(IF (EQ I9617 1 )(; (: MSK17 0 )(: JST17 0 ))(; (: MSK17
"4000 )(: JST17 1 )))(; (: MSK17 0 )(: JST17 0 ))))
PROGRAM-PATHS-PREDICATE
(( NOT ( KV96 EQ 1 ) ) AND ( NOT ( KV17 EQ 1 ) ) )
—
FALSE
—
(( NOT ( KV96 EQ 1 ) ) AND KV17 EQ 1 )
—
(KV96 EQ 1 AND ( NOT ( KV17 EQ 1 ) ) )
—
(KV96 EQ 1 AND KV17 EQ 1 )
—
FALSE
—
```

FIG. 3b PROGRAM CODE IN LIST FORM WITH RELATED PATH PREDICATES

MODULO ESAMINATO N. 71

MOD71:	KV17=	0	MSK17=	0	JST17=	0	
	KV96=	0					I9617= 0
MOD71:	KV17=	1	MSK17=	4000	JST17=	1	
	KV96=	0					I9617= 0
MOD71:	KV17=	0	MSK17=	0	JST17=	0	
	KV96=	1					I9617= 1
MOD71:	KV17=	1	MSK17=	0	JST17=	0	
	KV96=	1					I9617= 1

FIG. 3c PROGRAM RUN USING TEST DATA DERIVED SOLVING PATH PREDICATES

```
TABLE IS COMPLETE AND NON-REDUNDANT

                              1    2    3
I   1  KV17                   1    -    0
I   2  KV96                   0    1    0

A   1  BIT11 ALL.BIANCO       X    .    .
A   2  NESSUNA AZIONE         .    X    X

MISSING RULES

NONE


REDUNDANT RULES

NONE
```

FIG. 3d DECISION TABLE BUILT FROM PATH PREDICATES AND PROGRAM RUN RESULTS

```
E1=I1
E2=I2
E3=-E2
E4=E1.E3
E5=-E4
A1=E4
A2=E5
ACTIONS
A1   BIT11 ALL.BIANCO    ;
A2   NESSUNA AZIONE      ;
INPUTS
I1   KV17;
I2   KV96;
END
```

FIG. 3f INPUT TO CERLTAB DERIVED FROM ABOVE LOGIC DIAGRAM

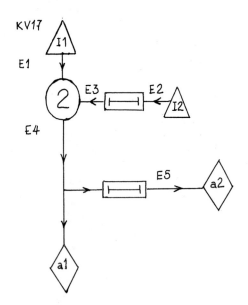

FIG. 3e LOGIC DIAGRAM DERIVED FROM SYSTEM REQUIREMENTS FOR OBJECT MODULE

```
TABLE IS COMPLETE AND NON-REDUNDANT

                              1    2    3
I   1  KV17                   1    -    0
I   2  KV96                   0    1    0

A   1  BIT11 ALL.BIANCO       X    .    .
A   2  NESSUNA AZIONE         .    X    X

MISSING RULES

NONE


REDUNDANT RULES

NONE
```

FIG. 3g DECISION TABLE CORRESPONDING TO ABOVE LOGIC DIAGRAM

Design and validation of software

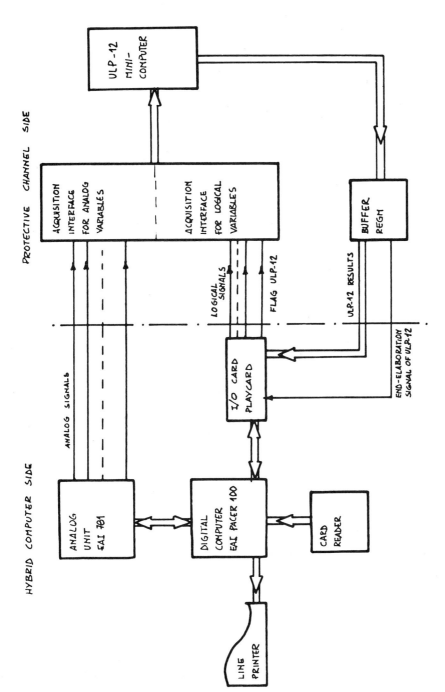

Fig.4 Hardware used in the System

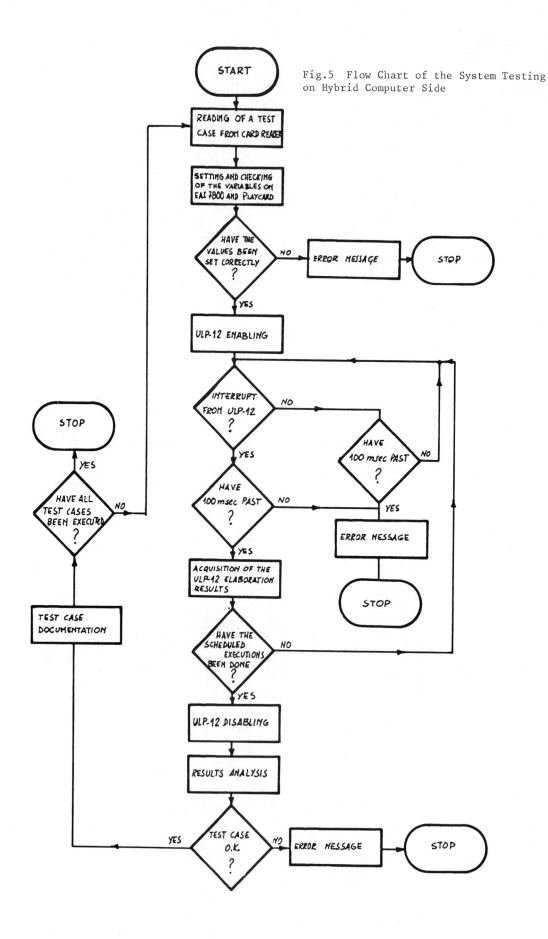

Fig.5 Flow Chart of the System Testing on Hybrid Computer Side

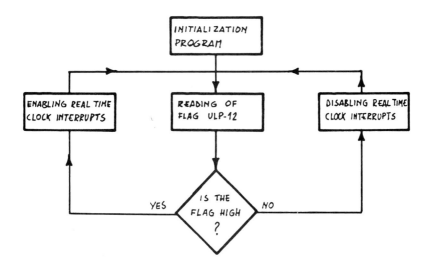

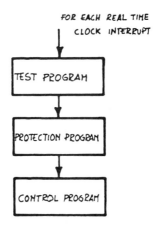

Fig.6 Flow Chart of System Testing on Protective Channel Side

GRAPHS OF DATA FLOW DEPENDENCIES

P. Puhr-Westerheide

*Gesellschaft für Reaktorsicherheit (GRS)mbH, Forschungsgelände,
8046 Garching, Federal Republic of Germany*

Abstract. Data flow dependencies of a program are important for code-optimisation of compilers and for detecting errors in programs. The control flow of a program can be described by its control flow graph. The mutual dependencies of variables in a program can be represented by graphs, too. For each variable of a program a tree structure can be found that shows its dependency on other variables. Different kinds of tree structures depict the data flow dependencies more or less detailed, according to the more or less exhaustive use of the source code's information. The most accurate representation of data flow by tree structures can be achieved for individual paths of a program only. A less accurate representation results from the set of unordered program statements. In this case it is not possible to distinguish whether a statement precedes another or not. By use of the static control flow graph, trees can be established without regard to branching conditions. With these trees distinctions can be made between different actual values of the same variable. The methods of building tree structures that show the data flow of programs are suitable for data flow analysis by hand as well as for automatic analysis.

Keywords. Automatic testing, data flow dependencies, graph theory, programming control, computer debugging, trees, analyzing tools.

1. INTRODUCTION

It is the task of software programs to control computers in such way that they provide meaningful output data dependent on the offered input data. Since computers are involved in processes that are either relevant for the safety of human beings and objects or where malfunctions would imply high economical losses, the intention arose to prove in advance the correctness of software programs for any combination of input data. Although it seems to be impossible to show the correctness of any arbitrary software program, several partial approaches to that aim were developed. Among these, one particular part is considered here: the field of reconsideration of a program's data flow. The instrumentation of programs allows to have a look at the flow of data during its execution, but it is also possible to get information about data flow by the static analysis of a program. Errors and anomalies like a missing initialisation of a variable, a reference of a variable that is undefined, or a definition of a variable that has not been referenced after its previous definition may be discovered, and possible decouplings between data may be detected. The mutual influences of control flow and data flow cause some unpleasent problems preventing a strict separation between them. Therefore, some special intersections between them, like considerations on path conditions or mapping conditions depending on paths are not treated in detail in this paper.

The analysis of data flow has its roots not only in the field of safety considerations, but also in the design of code optimizing compilers. The FORTRAN analysing system, for example, /1/ contains a data flow analysis routine.

2. SEVERAL METHODS FOR THE ANALYSIS OF DATA FLOW

A very qualified method to represent the mutual dependencies of program data is the method of the Symbolic Execution. Each path of a program can be executed symbolically, i.e., the values of input variables of a program are represented by symbolical values instead of numerical or other actual values. Along an interesting selected path, variable values and path conditions are collected and, as far as possible, repeatedly substituted by existing symbolic values or expressions until the final or any other interesting point of the path is reached. The values of variables and the path condition, then, is given by formulae depending on symbolic input variables only. By evaluating these formulae, the correctness of a path execution can be shown in relation to its specification.

The symbolic execution is a very powerful method for finding errors in programs /2/. It shows exactly the mapping of data of program runs. Unfortunately, since the final expressions are frequently rather complex, they are often unwieldy and, therefore, difficult to resolve and to understand.

The analysis of data flow due to the method of /3/ and others is a less sophisticated method to investigate data mappings, but able to depict a lot of data flow errors or anomalies when existing. This method has its roots in the design of code optimizing compilers. It is based on the premises that the last action performed on a variable is represented by one of the three states, namely defined, undefined, and referenced, and that the sequential arrangement of these states due to a program path permits to detect anomalies as, for example, the state 'undefined' followed by the state 'referenced'. That means, the attempt is revealed of reading the value of a variable that was not even initialized.

A modest representation of data dependencies is given by most compilers as cross reference lists. Some of them present two kinds of list elements pointing out whether a variable is defined or referenced.

3. ELEMENTS OF DATA FLOW GRAPHS

As the representation of the control flow by its graph, the mutual dependencies of variables can be depicted as graphs, too. These graphs are composed of elementary partial graphs that can be extracted from the assignment statements. Supposed all assignment statements have the following configuration or can be matched to it:

$$v_d = f(v_{r1}, v_{r2}, \ldots, v_{rn})$$

v_d ... d-variable (defined variable)
v_r ... r-variable (referenced variable)
f ... mapping

Let V be the set of all names of variables standing in the statements of a program. Within this set, the values of the constants c_i are regarded as their names and, in the following, marked by the letter 'C'.

Whenever a variable with a name in V (say A) maps its value to a variable with a name in V (say B) during the execution of a statement (for example: A:=B), an element (A,B) RcVxV is generated. The set RcVxV is a relation on V.   (1)
It can be established by a static analysis of the source code. Directed graphs are the elements of almost all data flow graphs mentioned in this paper[+]. The starting node of an elementary graph carries the name of a defined variable, its end node the name of a referenced variable. In the drawings the direction of edges is from top to bottom until otherwise noted. Statements containing more than one
+ except chap.5.4

referenced variable lead to more than one elementary graphs of one statement tree of a variable by merging the starting nodes of them, as shown in fig.1.

A statement tree of a variable contains the same information as all entries concerning one statement in a cross reference list. The statement data trees of a set of statements can be linked together in different ways, as shown below.

4. WORST CASE DATA TREES

Without analysis of the control flow, the set of statements is not ordered with regard to the sequence of execution. Hence, it cannot be decided which assignment to a variable was relevant or not after the execution of the program. Therefore, when ignoring the control flow, all assignment statements must be considered in the same manner, even if the last assignment possibly overrides all preceding assignments. The concatenation of all statement trees of a variable in a program by merging their root nodes creates a tree structure that depicts all variables to be referenced when defining the root variable in any statement of the program. It does not show indirect references via the dependencies of the referenced variables on further variables. This disadvantage can be avoided by linking the corresponding trees of the referenced variables to it. A problem may arise when a first variable is referenced during the definition of a second and vice versa[+]; in this case, the linking never ends. The agreement to link the tree of a variable only once removes this problem. The resulting tree is called a 'worst case static tree of implicit variable' WSTIVV, it shows all possible dependencies of a variable on other variables in a program. The fact that a variable is not contained in a WSTIVV means that it is decoupled from the root variable and may not influence the value of it by any assignment. (Note that it may influence the value of it via the selection of paths by branchings, and that the presence of a variable in a WSTIVV does not mean that a reference to the root variable exists mandatory.) Every variable depicted as a leaf in the tree must either be an input variable, or a constant[++], or it appears in another site as a node with successors; otherwise a reference of an uninitialized variable is detected. Fig.2 shows the construction of WSTIVVs for a program part.

A cross-reference-list whose list elements must be distinguishable between defined and referenced permits the construction of WSTIVVs.

[++] A constant is regarded as a special case of a variable (see also chap.3).

[+] Such a data graph is not a tree, as it contains cycles. However, by reasons of presentation, the cycles are cut off by splitting its start and end node into two nodes.

5. THE CONSTRUCTION OF DATA TREES WITH REGARD TO THE CONTROL FLOW GRAPH

5.1 Data Trees of Paths

The data mappings accomplished after a program run are related to the path that was run. A given path is a string of statements in the order of their execution. This information on the flow of control permits a decision on which of two statements is executed first, what was not possible when building WSTIVVs. In general, the selection of a path is dependent on the input data of a program. For that reason data trees built with regard to the execution sequence of a path are furthermore called dynamic trees of implicit variables of a variable (DTIVVs).

5.2 Data Trees Corresponding to a Whole Control Flow Graph

Another group of data trees with properties in between those of the DTIVVs and WSTIVVs does not need the consideration of actual values of input variables (i.e. the knowledge of a certain path) but of the static control flow graph. These trees are called STIVVs (static trees of implicit variables of a variable).

STIVVs of basic blocks. A basic block within a control flow graph "is a group of statements such that no transfer occurs into the group except to the first statement in that group, and once the first statement is executed, all statements in the group are executed sequentially" /4/. As described in chap.3, the trees of the defined variables are set up for every assignment statement within a basic block. After this step, the concatenation of the trees is following, beginning with the first statement of the basic block. (Now there is no need of merging the trees with the same roots). If any leaf of a new tree has the same name as the root of an already existing tree, it will be substituted by the already existing tree. After the concatenation process a tree with the same root that possibly exists already is erased.
In the final step of processing a basic block, each leaf of all existing trees gets a dummy successor as an interface node (except leaves representing constants). Fig.3 shows the building of STIVVs of a basic block.

Concatenation of STIVVs of basic blocks. A sequence of basic blocks of the control flow graph without any branchings is a part of a path or an entire path. For such a given path, the STIVVs of the basic blocks are concatenated by replacing the dummy nodes by the corresponding STIVVs of the preceding basic block. The results of concatenation are the DTIVVs showing exactly all data mappings in the same sequence as performed on a path. The treatment of all paths in this way mostly has its limits in the big number of paths; however, with a loss of information, another method of concatenation using the whole control flow graph by-passes this difficulty. There is no problem with a branching in the control flow graph where the succeeding basic blocks simply inherit the STIVVs of their (common) predecessor. But when a basic block, at a conjunction of partial paths, has more than one predecessor, it is generally impossible to decide which predecessor transfers the control to it with the knowledge of the static control flow graph only. Hence, the preceding basic blocks must be considered equally. Each variable within a basic block that inherits its value of preceding basic blocks gets as many dummy node successors in the STIVV as there are edges entering the basic block (fig.4). In the course of concatenation of the STIVVs each dummy node must be replaced by the corresponding STIVV of a previous basic block. In the final STIVVs of that kind, the dependency of a node variable on all of its successors is not mandatory for any given program run.

Concatenation of STIVVs in strongly connected regions of a control flow graph. A special kind of conjunction of partial paths is given in strongly connected regions of the control flow graph. A strongly connected region is a subgraph of a graph in which each node is both its own successor and its own predecessor. In a reducible flow graph (see /5/) (roughly spoken: each loop formation has only one entry node), each latching node that feeds back the control to the entry node of a strongly connected region is unique.

Fig.5 shows the scheme of a simple, strongly connected region. The concatenation of the STIVV of basic block 3 with those of basic block 1 yields the STIVV of the exit basic block (exit STIVV), containing[+] two nodes as dummies for the STIVV to be concatenated from the basic blocks that precede the entry basic block.

In the next steps the concatenation of STIVVs is continued until the STIVV of the latching basic block inherited the STIVV of the entry-basic block. Therewith the elementary STIVVs of the strongly connected region are found out. Since the number of runs through a strongly connected region is unknown in general, the presentation of data dependencies by STIVVs results in repetitions of its parts, as shown in fig.5. The repeated parts are the STIVVs of the latching basic block with the exception of their root node and that dummy node that represents the interface to the latching basic block. Fig.6 shows a strongly connected region with two latching basic blocks. Now three edges enter the entry basic block, two of them are latching edges. Since the control may be transferred either to basic block 3 or 4 each time when it is in basic block 2, it must be considered that the STIVV of the entry basic block inherit the STIVVs of basic block 3 and basic block 4 for each repetition. In this case, the repetitions in the STIVVs are twofold. For each strongly connected region a STIVV consists of a body and a repetition subgraph that is labelled and quoted only once in that STIVV.

[+] In case the corresponding root variable was defined within basic block 1 or 3.

5.3 Relations between Data Trees of Program Paths and of a Whole Program

A characteristic of the STIVVs of the last chapter is that they contain as subgraphs the corresponding data trees of each path included in the control flow graph (DTIVVs). They are the whole of all DTIVVs of a variable in a program. The selection of a special DTIVV by hand can be done by drawing a line that intersects those edges of a STIVV that belong to the DTIVV, in the sequence of the execution of assignments as shown in fig.7a.

To each edge of a STIVV the number of the corresponding statement can be attached. It facilitates the search for the corresponding assignment. Fig.7b shows a STIVV composed of nodes, edges, and a third kind of elements that combine all edges corresponding to a partial path entering a basic block in the control flow graph.

5.4 Array Elements in Data Trees

Array elements are handled as normal variables; their index variables are regarded as variables that are referenced during the definition of an array element. Although it is a simplification, it is in accordance with the practice in several compilers (fig.8).

6. DATA TREES OF AFFECTED DATA

The replacement of relation (1) in chap.3 by the following:
"Whenever a variable with a name in V (say A) maps its value to a variable with a name in V (say B) during the execution of a statement (for example: A:=B), an element (B,A) RcVxV is generated. The set RCVxV is a relation on V."(2)

enables the generation of data graphs representing the influence of a variable over other variables, the affected variables.

Fig.9 shows the worst case static trees of affected variables of a variable (WSTAVV) for a program part.

7. CONCLUDING REMARKS

The presentation of data flow dependencies as graphs is an easy comprehensible way to show the mappings of data in a program. It permits the finding of certain data flow anomalies as well as possible decouplings of data. Data trees composed of cross-reference-lists give a simple overview on data mappings that is not very distinct because of the neglect of the control flow graph.

A clearer insight is given by data trees corresponding to the whole control flow graph of a program. The most exact presentation comes from data trees of single paths, but it has the disadvantage to show only the mappings of a special path out of a number of paths which frequently is huge, preventing the analysis of the whole of paths by quantitative reasons.

The presented considerations of this paper on data trees were made in connection with the development of an analysing system for program modules written in PEARL, a project that is supported by the BMFT[+]. Since there are so many relations between the methods of compiling and analyzing programs, the decision was made to choose an intermediate language as starting point for the analysis of modules. A particular PEARL-compiler translates PEARL-programs into strings of an intermediate language. The analyser is scheduled for carrying out some static analytic tasks, beginning with the structural analysis (see/6/) of the intermediate strings. It should become a tool that is particularly helpful for safety assessment. In connection with the analyzer a set of rules will be provided whose observation during program development will facilitate the analysis. Particular care is taken to provide a convenient presentation of the analysis results.

I would like to thank PDV (Project Prozeßlenkung mit DV-Anlagen) for its support and messrs. Grumbach and Märtz for reading the manuscript.

8. REFERENCES

/1/ Fosdick, L.D., and L.J. Osterweil.(1975) DAVE - A Fortran Program Analysis System. Dept. of Computer Science, Univ. of Colorado, Boulder, Colorado 80302.

/2/ King, J.C. (July 1976). Symbolic Execution and Program Testing. CACM vol.18 no.7.

/3/ Fosdick, L.D., and L.J. Osterweil (1976). Data Flow Analysis in Software Reliability. Computing Surveys, vol.8, no.3.

/4/ Aho, A.V., and J.D. Ullmann (1973). The Theory of Parsing. Translation and Compiling, vol.2, Poentice Hall, Englewood Cliffs, N.J., p.912.

/5/ Hecht, M.S., and J.D. Ullmann (Oct.1973). Analysis of a Simple Algorithm for Global Flow Problems. Conf. Record, ACM Symp.on Principles of Programming Languages, Boston, Mass., pp. 207-217.

/6/ Okroy, K. (June 1978). PDV-Bericht KfK-PDV 152, KfK Karlsruhe.

DISCUSSION

Grams: What types of programming errors can be detected by your method?

Puhr-Westerheide: For example variables which are not initialized or not referenced.

Maki: What is the advantage of this technique? A compiler can also find for example not initialized variables.

Puhr-Westerheide: Yes, you can also build the worst case data trees starting from the cross-reference-list of a compiler too.

+ BMFT ... Bundesministerium für Forschung und Technologie

A:=A+C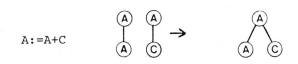

assignement   tree elements   assignement variable tree

fig. 1. Merging the elementary graphs of an assignement statement

A:=D+E+F
E:=B+D

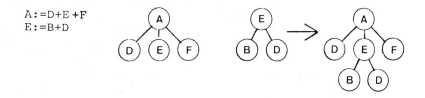

fig. 2. concatenation of STIVV's

Maki: So why do you use your technique?

Puhr-Westerheide: First: You can see which variables may affect a variable which is denoted in the root-variable. If you do this with the cross-reference-list of a compiler it is a very hard work.

Second: The information of the control flow graph is also used. You can also set up a very exact and clear data tree for only one path which shows exactly all mappings on one path. This cannot be done by a compiler.

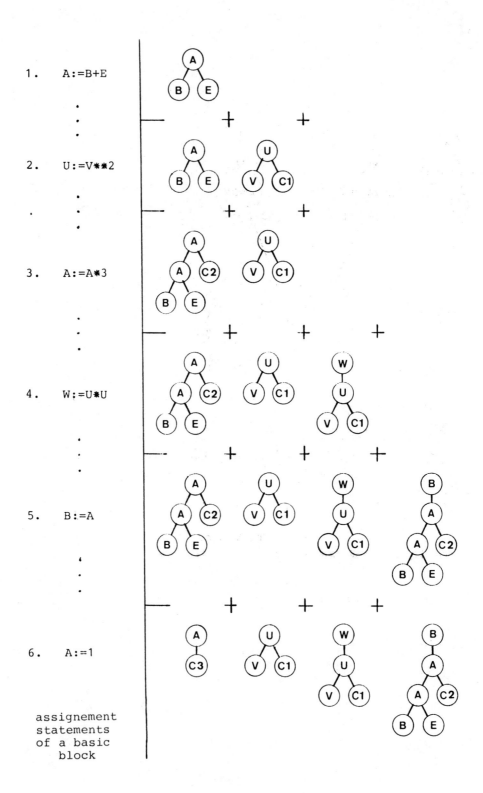

fig. 3. The generation of STIVV's of a basic block

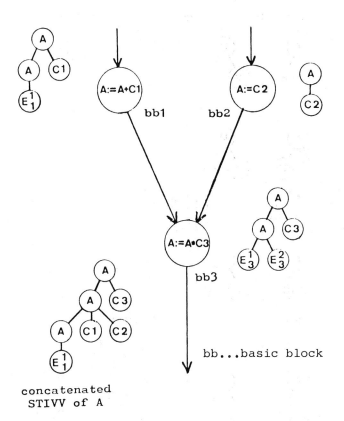

fig. 4. Concatenating the STIVV's of basic blocks

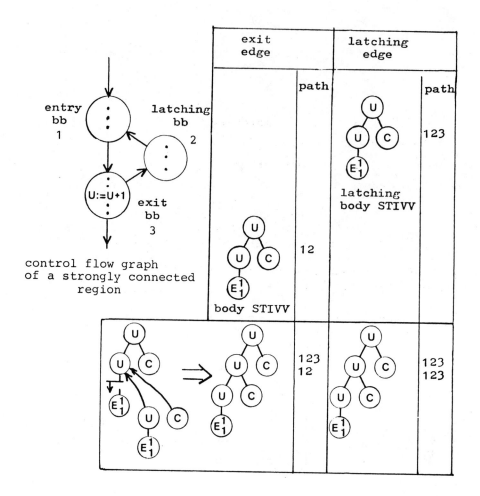

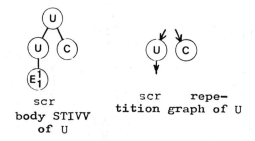

fig. 5. The generation of STIVV's of a <u>s</u>trongly <u>c</u>onnected <u>r</u>egion (scr)

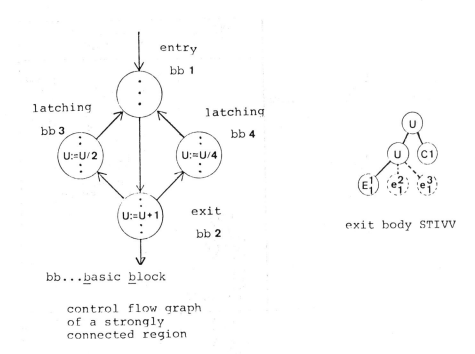
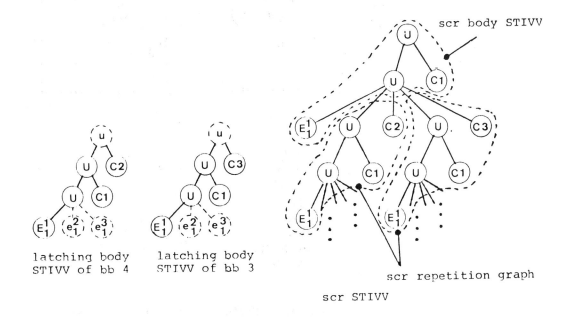

fig. 6. The generation of STIVV's of a scr containing two latching basic blocks

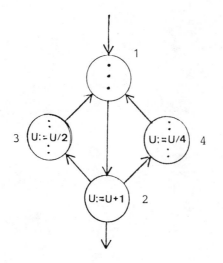
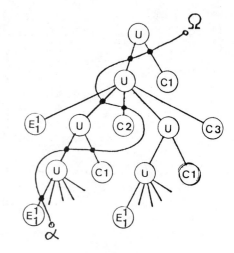

control flow graph

path 12312

α begin of the path
Ω end of the path

STIVV

fig. 7a. A line labels a subgraph corresponding to a path

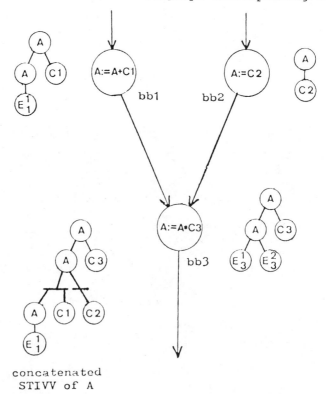

concatenated
STIVV of A

fig. 7b. STIVV of fig. 4, extended by a new element (see chap. 5.3)

A(I):=A(I-1)+A(I)

Fig. 8. STIVV of an array variable

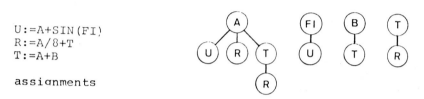

```
U:=A+SIN(FI)
R:=A/8+T
T:=A+B
```

assignments

fig. 9. Trees of affected variables of a program part

SAFETY PROGRAM VALIDATION BY MEANS OF CONTROL CHECKING

W. Ehrenberger and S. Bologna*

*Gesellschaft für Reaktorsicherheit (GRS)mbH, Garching,
Federal Republic of Germany*
*Comitato Nazionale Energia Nucleare, Casaccia, Italy

Abstract. Recent progress in the field of computer hardware makes reasonable the search for improvements in software reliability by using larger and faster computers, which permit the usage of redundant programs or the application of redundancy during the execution. The basic idea is to memorize what has been tested during the debugging, licensing or burn-in phase of software and to switch to the safe side if any program status appears during on line operation that had not been tested before.

Concerning control sequence checking, it is supervised during program execution whether the program path used has already been tested. Three alternatives are considered: The first stores the tested paths as a tree, which describes the connections between the consecutive arcs. The second one associates an identification number to each program arc and stores the string of numbers, that identify each tested path. The third maps each program path on a number and stores it.

In the same way also three checking methods for addressing arrays are feasible: The first memorizes the tested sequences of addressing as a tree structure, the second memorizes the tested sequences of addressing as a string of numbers and the third maps the sequence of addressing, used during a program run, on a single number.

The size of the overhead in program execution time and memory space required is a function of the paths and mappings memorized. The application of one or more of these methods seems reasonable primarily for safety relevant programs which have to meet nearly identical requirements for long periods of operation time, and where safety actions are normally appropriate if input conditions change to unanticipated constellations.

Keywords. Software Reliability, Testing, Memorizing Test Cases, Software Redundancy.

INTRODUCTION

Concerning the currently available techniques and tools for software design and validation some limitations continuously exist. The major ones are /1/:

- The existing validation methods based on analysis of the internal structure of the program sometimes are not applicable for complicated programs, or are extremely costly.

- No existing method can guarantee the correct execution of a program by the computer hardware, even if the program was a-priori logically correct.

- Unexpected data resulting from environmental abnormalities can interfere with the execution of an otherwise correct program.

Thus the idea was born to add some redundancy to an existing program which improves the probability to get a correct execution when the program is used on line.

For the following we always assume to have a technical process which has a safe side. Such processes are for instance railway systems and nuclear power plants. The method to be described later checks whether the actual execution of that program has already been tested. If it detects that it has not been tested, an action to the safe side is initiated.

Basically all the techniques to recognize an abnormal behaviour of the system can be classified according to three aspects /2/.

They can check

- the mappings performed by the program
- the control flow through the porgram
- the data handled by the program.

Checking the mappings of the program means to check the plausibility of the outputs for a given set of input data. Checking the control flow is to check whether the sequence of computations executed was expected or not. The aim of data checking is to assure the integrity of a data value, the integrity of a data structure, and the nature of a data value.

In this paper we deal with control flow and data flow checking. As we will see later, the approach proposed is rather time and memory consuming. Since, however, due to the progress in computer hardware, computing time and computer memory get cheaper and cheaper, we think it is usable for future computer systems.

In the literature several techniques are suggested for the detection of an error in the sequence of computations /2,3,4,5,9/. They are all based on the nature of the error we may expect to have, for instance infinite loop, illegal branch, etc. and they imply to design software such that the program checks its own dynamic behaviour automatically during execution. On the other hand, the approach we suggest is based on the results obtained during the program testing phase.

THE PROPOSAL

If we think in terms of a program control graph, every sequence of computations during a program run corresponds to a path through the program from the entry node to the exit node. The approach described in chapter 4 is based on observing and memorizing which paths have been tested. If during on-line program execution a path is traversed which has not been tested before, it is assumed that the result of its mappings may be false. In the same way as we can think of a program control flow graph, we can think of a data flow graph. The approach described later observes and memorizes data movements during the test phase and monitors during the execution phase whether untested data paths are used. What we propose can be summarized in the following three points:

a) The source program is instrumented such that it is able to memorize each control path and each data path executed during the test phase.

b) The program is tested exhaustively according to the information from its specification, from its internal structure, and according to the general knowledge about the problem to be solved. Each tested control path and data path is carefully watched whether it leads to correct results, and then is memorized.

c) During on line operation, each path traversed is checked against the paths memorized during test phase, to see if it was expected or not. If we have an unexpected path, this can be due to:

1) The program is correct, but it was executed incorrectly (execution error, e.g. due to sporadic hardware fault).

2) The program is correct and is executed correctly, but was supplied with a new input condition, which lies outside the specified range and was not considered during testing (no error).

3) The program is not correct and conseuqently executed incorrectly because the supplied new input condition was never thought of before, though it lies in the specified range (specification error).

4) The program was executed incorrectly although input data lie in the specified range (programming error).

Obviously only case 2 is not dangerous, though this may happen frequently due to non-exhaustive testing. In our proposal always the worst cases, cases 3 or 4, are assumed and a reaction to the safe side is suggested.

Fig.1 shows the general organisation of the suggested program testing and run strategy.

THE PROGRAM INSTRUMENTATION PROBLEM

In general the question of finding suitable locations for monitors as well as inserting them into a selected location is a difficult problem. In order to avoid errors, the insertion should be suitable for automatic execution. In the literature several methods are suggested concerning both the optimal placement of monitors, and the results which should be memorized. In general, the more monitors are employed, the more accuracy or information can be obtained /6,10/. However, each monitor results in an operation overhead of execution time and storage. This makes it desirable to install a minimum numbers of monitors at those locations in the target program which provide all essential information requested. In our case the information requested must be sufficient for an unambiguous identification of the control and data paths traversed.

For the control paths it is sufficient to choose, as program element to be instrumented, the so called program arc, that is a (possibly empty) sequence of executable statements between two structure elements of a program. Each arc begins at one specific exit of a structure element and ends at the next structure element. Structure elements are for example IF THEN ... ELSE, CASE, WHILE-DO.

flow, if the indices depend on input data via mappings and not only via control structures. We therefore restrict our considerations to array addressing in addition to control flow supervision.

In general two approaches to identify the traversal of a still untested path are conceivable. One is to detect the traversal of the path as soon as possible by using a technique which we call "following the program tree" or "following the data tree". The other is to detect such a situation at the end of the program run by using the technique we call "associating an identification number to each arc in the program" or the technique "mapping each path on a number"; the data flow analogies are "associating an identification number to each array addressing" and "mapping each data path on a number".

In both approaches the major task is performed by two routines. They are called Test Supervising Routine and On-Line Supervising Routine. The first one memorizes the data acquired during the test phase, the second one performs the comparison during on line execution. These routines vary from method to method.

In order to avoid that the error detection code may be damaged or bypassed when an error occurs, allowing the error to remain undetected, it is important to protect the integrity of the error detection code. The Test Supervising Routine and the On-Line Supervising Routine must both have the capability to ensure their own integrity.

CONTROL SEQUENCE CHECKING

Following the Program Tree

The Instrumentation

The instrumentation consists of inserting calls of a specific subroutine at every outcome of a structure element, or thinking in terms of a program graph, at every arc. The parameter of the call identifies:

a) Which kind of structure element has been reached immediately before the instrumented point was reached.

b) Which one of the acrs starting from the structure element was used.

The Test

By means of the described instrumentation each traversed path can be identified unambiguously and memorized in a tree from data structure like in Fig. 2 during the test phase. A structure elment corresponds to each node of the tree.

The IF-THEN-ELSE element is described by two components containing the pointers to the subtrees corresponding to "condition satisfied" and "not satisfied". The CASE element is described by a number of components corresponding to the acrs departing from the

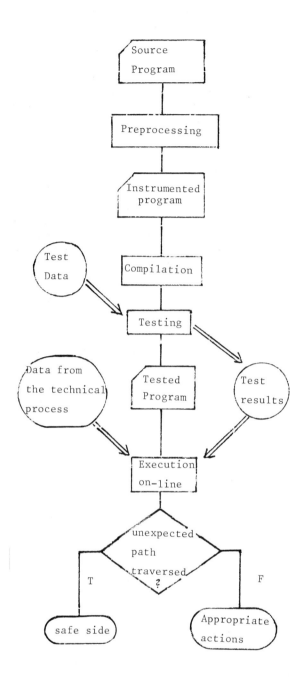

Fig. 1: General organisation of the program testing and run strategy

For a data path it is suffficient to instrument the program such that any addressing of a data location in the main memory area related to the program is memorized when the corresponding parameter is used. In most cases, however, the supervision of such data addressings is redundant to the control flow: if a special control path is traversed, the addressed data are fixed, i.e. they do not vary during that path. Then data flow supervision would cause additional costs but would not increase our knowlwdge about program execution. There is one exeption to this: the addressing in arrays. This addressing is characterized by its independency of control structure, each of which contains a pointer to the related sub-tree.

The WHILE-DO element is described by two components containing the pointers to the sub-trees corresponding to "loop to be repeated" and "not to be repeated". If we have a loop with a variable, input depending number of repetitions, one additional component is necessary for the pointer to a node representing the same structure element. During each new repetition of the WHILE-DO construction, a new corresponding element is initiated. If we have only a "simple loop", i.e. no other structure elements are contained in the body of the loop, the component containing the pointer to the sub-tree corresponding to the inner loop code will contain only information on the number of runs through the loop.

For every element the value assumed from the pointer is NIL when the corresponding sub-tree is empty. In this way each traversed path in the program is unambiguously identified by a path through the tree. The tree with its components is constructed by the Test Supervising Routine.

The Execution Phase

At each instrumentation point it is checked whether the corresponding arc has already been used in the context of the present execution. This is done by the On Line Supervising Routine.

Discussion

This technique requires a subroutine call during the execution of each arc. This is very time consuming. Therefore it is not suitable for programs with many arcs. The overhead for the execution time and for the memory space needed is a function of the density of structure elements in a program.

The technique shows whether the corresponding path has already been tested at each instant of the execution of the path. Therefore, it is suitable for programs where the execution should be stopped, once the flow gets out of the corresponding tested path. Since at the entering point of any arc, it is checked whether that arc had been tested before in that path, no untested path will be able to produce uncontrolled output.

Associating an Identification Number to Each Arc

A slightly different way can be applied to use the information stored during the test phase in order to detect unintended paths during program run. Here, the executed path is recorded and a comparison is made between this path and memorized paths at the end of the program run. This is possible by associating to each arc in the program an identification number and to characterize every program path by a string of identification numbers.

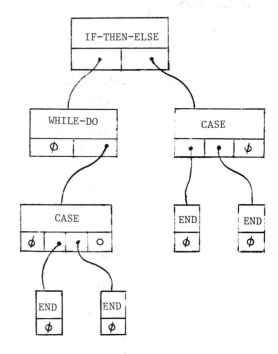

Fig. 2a: Tree control structure, possible result of 4 runs of a program like fig.2b

The Instrumentation

Each arc of the program is instrumented by inserting a software probe which consists of a special auditing subroutine. The function of the auditing subroutine is simply to concatenate the identification number of the executed arc to the string of identification numbers stored in the reserved storages that identify a particular path through the program. Each time the probe on arc 1 is tripped a new test case begins.

Test and Execution Phase

During the test phase each tested program path provides a string of identification numbers which identify in unambiguous manner the path traversed (Fig.3). A call to the Test Supervising Routine is inserted just before the end of the program in order to store all the different strings of identification numbers produced during the test phase.

During the execution phase the call to the Test Supervising Routine is replaced by a call to an On-line Supervising Routine which checks whether the string of identification numbers representing the path is already in the store.

Discussion

Given that every path is characterized by an unique string of identification numbers, the number of comparisons that must be made in order to identify that a path has already been executed reduces itself as the string is

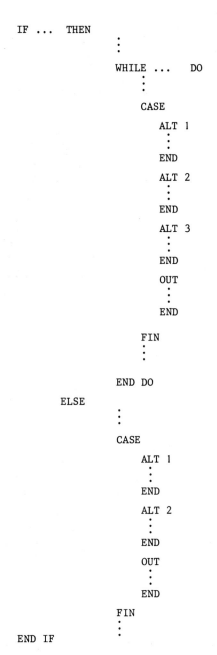

Fig. 2b: Program corresponding to tree control structure of Fig. 2a

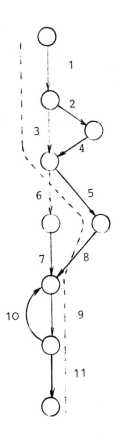

dotted path: edges 1,3,5,8,9,11

Fig. 3: An Identification Number is Associated to Each Arc

searched. Let N be the number of tested and stored paths of the program: For the first identification number in the string we then need N comparisons, for the last identification number in the string we ideally need 1 comparison. As a consequence the expected number of comparisons can be estimated from the triangular distribution. For example, if the object path has 200 arcs and the number N of tested and stored paths is 1000, the number of comparisons that we expect to carry out is given by:

$$n = \frac{1000 \cdot 200}{2} = 100\ 000.$$

Mapping Each Control Path on a Number

The two techniques described before need much memory space and may be very time consuming. The following technique uses information evaluated on each executed arc, but stores information only on each executed path.

The Instrumentation

Each arc of a program is instrumented by one additional simple operation.

Operator $\rho \in \{+,-,\times,/\}$. Operand $k \in \{\text{prime numbers}\}$

$k_{\text{arc } i} \neq k_{\text{arc } j}$, $i \neq j$. The value assumed from a variable, say A, is to be used to identify a particular path.

$A_i := A_o \rho_1 k_1 \rho_2 k_2 \ldots \rho_n k_n$, path i with n arcs.

Test and Execution Phase

During the test phase each program path provides one particular value of A. A call to the related Test Supervising Routine is inserted just before the end of the program in order to store all the different values assumed from A. During the execution phase the call to the Test Supervising Routine is replaced by a call to an On-line Supervising Routine which checks whether the value of A gained by the executed path is already in the store. If yes, the program reaction, which was the result of the path's calculation is initiated; if no, a reaction to the safe side is activated.

An example is shown in Fig.4.

The Problem of Ambiguity

In contrast to both the methods mentioned before, this approach does not lead to unambiguous results: It may happen that an untested path leads to the same characteristic number A, as a tested path. This does not necessarily imply an error in the corresponding result of the computation. The untested path can have been executed with no error. The probability for this can be estimated as follows:

n number of tests performed

n_e number of erroneous runs performed

$\frac{n_e}{n}$ relative frequency of erroneous runs

- = estimate for the relative number of erroneous runs to be taken into account for the untested paths,
- = estimate for the probability of getting an error during the execution of an untested path.

In order to be able to use the above consideration, we must assure that

a) no error is corrected during the n runs

b) the n runs have been organised such that the probability to find errors was equal for each run and independent from all other runs.

Discussions

This technique is applicable if the supervised program generates its output at its end and clearly separated from the part where the results are evaluated.

The overhead for cpu-time and memory space needed is less than for the other techniques.

If the program contains many loops, or loops with large repetition numbers, it is necessary to check whether the limits of the representable number range are not surpassed.

DATA FLOW CHECKING

The principles of data flow checking shall not be treated so detailed as the principles of control flow checking. We try to be rather short here, since nothing is essentially new. As already mentioned, we only deal with addressing in arrays. We further assume that control flow checking is applied in the investigated program and we therefore relate our memorizing of data paths to already memorized control paths. That means, if an already stored data path is executed in connection with another control path, it is memorized a second time in connection with that second control path. During on line execution it is first searched whether the traversed control path has been tested and then whether the related data path has been tested.

Similarly to the control flow supervision we have here three methods to discuss.

Following the Data Tree

Each already used data path is identified unambiguously and memorized in a tree-form data structure during the test phase. To each node of the tree corresponds the address of an array element. In that tree each array element succeeds the element that has been used immediately before. It is in turn succeeded by those elements that are used immediately after.

Associating an Identification Number

In connection with each control path a sequence of numbers is stored, characterizing the related data paths by describing the different sequences of array addressing that occured during that control path. Each number corresponds to a particular array addressing, marking both the array and the position in the array.

Mapping on a Number

The numbers received are not stored, but connected with arithmetic operators. Only the ultimate result is stored in connection with the number characterizing the related control path.

In another version control path and data path number could also be connected and stored as one parameter, thus saving one storage place and one comparison per control path.

GENERAL REMARKS

Use of the Suggested Checkings

The goodness of the suggested approach depends very much on the goodness of the data stored during the testing phase. I.e. it is valuable if the data used during testing are similar to the data coming from on line demands, and it is useless if not. The test result storing phase should start only when the program was already completely tested and the programmer has had the feeling that it is running correctly. All test runs must be carefully watched whether they lead to correct results. If an unexpected result is observed, the test must be stopped and the program corrected. After

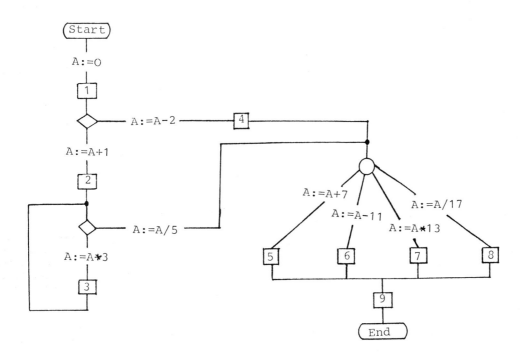

Fig. 4a: Instrumented program according to the strategy of mapping each path on a number

Path	sequence of program arcs	value of A at point END
1	1 2 3 (10 times) 7 9	15 3527.4
2	1 4 8 9	-0.1176471
3	1 4 5 9	5
4	1 2 3 (4 times) 6 9	70
5	1 2 3 (4 times) 7 9	1053

Fig. 4b: Paths from Fig. 4a and the corresponding value of A at point END

correction of the program the test phase is to be started again from the beginning.

Limitations

None or our techniques checks errors coming from numerical calculation deficiencies due to insufficient algorithms or inappropriate computer word length. In particular we can not find out whether the used algorithmus are suitable over the whole range of the corresponding input domain.

Since it is the control paths we are dealing with, we also do not check whether a path is related to the correct predicate. That is, if the path is executed correctly, but is related to a wrong subdomain of the input data, we will not necessarily find the error. Thus it is impossible to find programming errors resulting in incorrect boundaries between different subdomains.

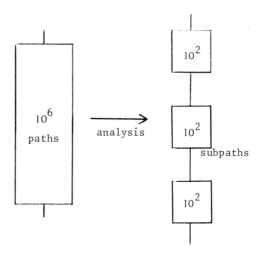

Fig. 5: Reduction of the number of paths to be considered from 10^6 to $3 \cdot 10^2$

Experience, however, shows that a considerable number of programming errors are errors in the control flow of the program. E.g., /8/ reports that the number of errors found during the test phase of some programs is quite well correlated with the number of branches of these programs.

Last not least our methods are related only to sequential programs. They do not deal with the correctness of interrupt handling, queuing or access to common resources.

This paper has tried to bring a new approach to on-line program self-supervision. It is certainly possible to improve and refine the methods. Also a combination of the described techniques together with other plausibility check techniques could be considered. However, we must take into account that the overhead of the error detection procedures is proportional to the complexity of the error detection procedures themselves and a greater reliability can be achieved only at the expense of greater overhead.

Modularization and its Impact

Taking into account that the test effort, the memory space and the execution time needed increase linearly with the number of paths to be supervised, one can try to reduce this number by means of modularization. If e.g. the program to be supervised has 10^6 relevant paths, it does not seem feasible to test and memorize each of them. If we can, however, split this program into three building blocks which work completely independently, it may be sufficient to supervise only 10^2 paths in each block. In some cases it is reasonable to make no instrumentation at all. Such parts of software could e.g. be

- mathematical standard routines, because they have been used very frequently and it is known from long experience that they perform correctly,

- simple logic decision routines, e.g. performing two-out-of-three decisions, because they have been completely analyzed and all possible combinations of input data have already been offered, thus having executed an exhaustive test,

- routines, whose correctness has been proved formally.

In these cases it could, however, be advisable to introduce some checkings, consuming less time and memory, which compare the number of loop executions against an upper bound or which verify the plausibility of the data (if they are within the definition range). If for any part of the software it is known in advance that it need not fulfill safety requirements, it can also be left uninstrumented.

Possible Applications

As already mentioned, all our techniques lead to a considerable overhead in computing time and memory space. Thus they are intended for applications where these aspects are of minor importance and where the safety of the program is essential. Areas of this kind are

- supervision of railway systems
- air craft control
- nuclear reactor safety.

Our approach qualifies above all for the last item. During the life time of a nuclear power plant, usually several hundred reactor scrams may occur. To a large degree they are initiated by a type of input conditions that is similar from case to case. Therefore it appears feasible to approach the advance testing of virtually all input conditions that may occur during the plant's life time.

The suggested approach can in particular be used in connection with software licensing: During the verification and licensing phase of a software package, all tested cases are memorized; if during the application phase an input state occurs that had not explicitly been licensed, an action to the safe side is initiated.

PRELIMINARY RESULTS

The suggested approach has been applied to some different modules of an already existing reactor protection software package. This package was developed by Gesellschaft für Reaktorsicherheit some years ago and has been used in the test application of a computer based reactor protection system.

From the collected data it may be concluded in general: that the overhead memory due to the instrumentation is insignificant with respect to the extra memory necessary to store the tested paths. The size of the overhead memory is a function of the number of paths and, at least for the first two techniques, the length of the paths. The overhead time due to the instrumentation is a function of the number of stored paths, at least for the last two techniques.

Until now no concrete results are available on the quality of the suggested approach with respect to on line operation.

Experience is continuously gained, however, and there has been no indication yet that the method cannot be applied.

REFERENCES

/1/ Hecht H. (Dec.1976). Fault-Tolerant Software for Real-Time Applications. ACM Computing Surveys, vol.8, No.4.

/2/ Yau, S.S., and R.C. Cheung (Apr.1975). Design of Self-Checking Software. Proceedings of International Conference on Reliable Software.

/3/ Kane, J.R., and S.S. Yau (March 1975). Concurrent Software Fault Detection. IEEE Trans. Soft. Eng., Vol. SE-1, No.1.

/4/ Bologna, S., and A. Daoud (14.-17.Sept.76). Defensive Techniques Increase Software Reliability. 3rd Congress National de Fiabilité, Perros-Guirec, France.

/5/ Ramamoorthy, C.V., R.C. Cheung, and K.H.Kim (1974). Reliability and Integrity of Large Computer Programs. Computer Systems Reliability, Infotech State of the Art Report 20.

/6/ Ramamoorthy, C.V., K.H. Kim, and W.T. Chen (Dec.1975). Optimal Placement of Software Monitors Aiding Systematic Testing. IEEE Trans. Soft. Eng. vol. SE-1, No.4.

/7/ Nickl, E., and M. Masur (Aug.1978). Erprobung einer Kontrollfluß-Überwachungsmethode anhand zweier FORTRAN Routinen. Gesellschaft für Reaktorsicherheit, Interner Bericht I-29.

/8/ Akijama, F. (1971). An Example of Software System Debugging. Proceedings IFIP-Congress. Lubljana.

/9/ Yau, S.S., R.C. Cheung, and D.C. Cochrane (1976). An Approach to Error Resistant Software Design. Proc. of the 2nd Int.Conf. on Software Engineering, IEEE Cat.No. 76CH1125-4C.

/10/ Stucki, L.G. (1973). Automatic Generation of Self-Metric Software. Proceedings of IEEE Symposium on Computer Software Reliability IEEE Cat.No. CHO 741-9CSR.

DISCUSSION:

Aitken: There is a very large number of influences to the software in a safety related computer system. Have you checked out this state of affairs in any way?

Ehrenberger: Definitely not. This is a work which still to be done.

Quirk: Do you count the loops in a program?

Ehrenberger: We did not intrument the loops explicitly, we instrumented all branches. That includes the loop-branches. As you mentioned the loop repetitions are more or less counted in our last method. In the other methods the whole supervision procedure is also repeated, if the loop is entered again. This leads to large overhead in main memory and computing time.

A PROCESS COMPUTER FOR EXPERIMENTAL USE

H. Holt

Laboratory of Automatic Control Systems, Technical University of Denmark, B.327, 2800 Lyngby, Denmark

Abstract. Current microprocessor based process computers uses read only memory (ROM) as program memory. Flexibility is obtained using interpreter like software. A process computer for experimental use must hold greater flexibility. Programs and subroutines in high level language must be able to run on the computer. There is a need for bulk storage like a floppy disc for program storage and data logging. The plant to be controlled must receive suitable stimulation under all conditions.

These characteristics are obtained by using a standard microprocessor and equiping it with suitable error detecting and handling hardware. It will be shown how to "control" the microprocessor from Texas Instruments TMS 9900.

Keywords. Computer architecture; computer control; computer hardware; direct digital control; industrial control; microprocessors; multiprocessing systems; parallel access storage; parallel processing; process control.

INTRODUCTION

This paper describes the features and design considerations for a microprocessor based process computer. The process computer is intended to demonstrate novel control algorithm performance compared with wellknown practice and is to be used in industrial environments and also as an educational tool.

The safety of a process computer is based on a hardware as well as a software reliability. The hardware used in microprocessor systems have proven very high reliability. The software reliability is dependent on the implementation. If the software has the form of a fixed program in read only memory, the software can be checked prior to installation on the plant. If some error occurs, it will always be possible to start from scratch, since the program memory always contains the correct instruction sequences. If the software is in read/write memory, one software error can modify memory content and thereby inhibit normal operation. On a process computer for experimental use, there is a need for interactive installation of tasks, and thereby for programs in read/write memory.

The interactive installation of tasks is performed by a user as a result of his needs for data manipulation, data logging or a change of a regulator algorithm. The problem is now, how to secure tasks already in memory, from software errors performed by a user-installed task. It will be shown in the following text, how to handle all possible user software errors.

The process computer is composed of a process controller and one or more process interface units. The process controller and the process interface units communicate via a standard terminal interface (V24/RS232C). The process controller is purely digital and is built as a multiprocessor system as described in the following chapter. The safety of the process controller is based on the processor board, which can run in either of two modes (monitor/user) and a well working operating system running in monitor mode. User programs run in usermode cannot modify or interfere with other user programs or the operating system.

The process interface is built as a separate microcomputer system with analog and digital links to the physical plant.

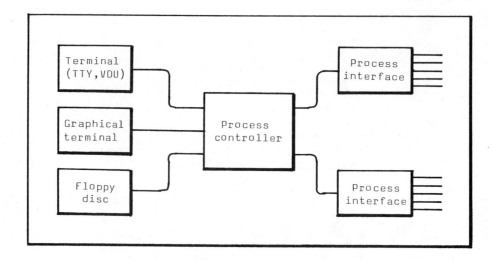

Fig. 1. Process computer outline.

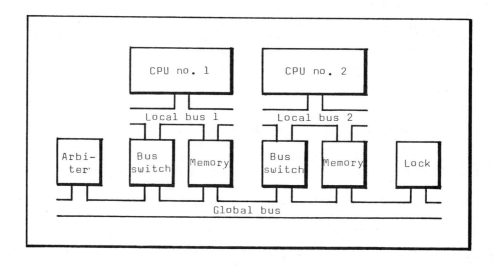

Fig. 2. Multiprocessor structure.

THE MULTIPROCESSOR STRUCTURE

This chapter describes the structure of the process controller. The process controller is implemented as a multiprocessor system. A real time operating system is supported by a lock-module and a notify mechanism.

Due to the lower capacity of a microprocessor compared to that of minicomputers and because the application requires different tasks to run simultaneously, it was decided to use the microprocessor in a multiprocessor environment. The basic design objective of the multiprocessor structure is that each processor runs different tasks, thereby keeping interprocessor communication at a very low level of activity. Efficient communication capability is provided for anyway, enabling high speed data fetching and fault recovery schemes.

Multiprocessor Busses

The microprocessor TMS 9900 uses several memory requests on each instruction due to its memory-to-memory architecture. This means that the microprocessor must directly access memory used mostly by itself, in order to minimize memory fetch delays. On this background the multiprocessor structure was based on a local bus for each processor and one global bus.

Thus all memory is designed as dual-port modules. One port is used for the global bus and the other is used for one of the local busses. This means that all memory in the system is accessible via the global bus and that each processor has access to memory without waiting for access to the global bus. If a processor addresses a memory module not existing on the processors local bus, the busswitch between this local bus and the global bus is automatically activated. When a busswitch is activated it waits for access to the global bus after issuing a bus request. Busrequests from all busswitches are examined by an arbiter. The arbiter assigns the global bus to one of the busswitches waiting for access to the global bus. When the memorycycle is finished, the busswitch releases the global bus. Each processor cannot have access to the global bus for more than one memory cycle in succession, thereby the global bus is only blocked when a processor uses it.

Lock and Notify Concepts

In a multiprocessor system there is a need for a synchronization function. The synchronization function is used to exclude simultaneous access to common data. The synchronization function is based on a hardware synchronization and cannot be simultaneously executed by different processors. The synchronization function consists of two subfunctions, named "lock" and "unlock". The lock function is a "test and set to one" of a memory register, the function being executed as a non separable element. The unlock function is based on a memory register clear (set to zero). When the microprocessor TMS 9900 is used in the described multiprocessor configuration, it does not support an instruction for the lock function, but of course for the unlock. A special memory unit named "lock-module" is therefore included, in which a memory read sets the memory content to one and transfers the original content to the processor (viz. either one or zero). The unlock function is performed by a memory write, which sets the memory content to zero.

The lock function is used to secure basic operating systems data from being accessed simultaneously by more than one processor. By using several lock variables, it is possible to lock minor blocks separately. This will allow different processors to access independent operating systems data simultaneously.

The lock function is also used when setting up a block of data for interprocessor communication. When the data block is ready for transmission the data block is unlocked and the destination processor is notified. This notification can be realized in different ways. The cheapest solution is to use a memory register as a flag, then the destination processor has to test the flag from time to time. This solution gives long response times. Another solution is to implement the notify function by an interprocessor interrupt, either separately or on a common notify line. The interrupt implemented notify function has a fast response time. We use the interrupt implemented notify function, but this will not exclude using the flag concept.

The notify function is also needed, when an operating system function run on one processor must have some part of the function executed by another processor.

THE MICROPROCESSOR
TMS 9900

This chapter will describe shortly the used microprocessor, as an introduction to the error detecting and handling described in the next chapter.

The microprocessor TMS 9900 from Texas Instruments is a single-chip 16-bit central processing unit (CPU). This microprocessor came on the market in 1976. The key features for this microprocessor is shown in TABLE 1. Among these key features, the memory-to-memory architecture and the I/O-bus structure is unique in single-chip microprocessors. The microprocessor has instructions for multiplying and dividing, features appearing only on novel chips, now becoming available. At the time of system definition, therefore, the TMS 9900 chip was a natural choice.

Memory-to-Memory Architecture

The microprocessor employs a memory-to-memory architecture. Blocks of memory designated as workspace replace internal-hardware registers with program-data registers. The microprocessor has three internal registers accessible to the user:

    PC - program counter
    ST - status register
    WP - workspace pointer

TABLE 1 TMS 9900 Key Features

1. 16-bit instruction word
2. Full minicomputer instruction set
3. 64 k bytes of memory
4. 3.3 MHz speed
5. Memory-to-memory architecture
6. Separate memory, I/O and interrupt busses
7. 16 general registers, residing in memory
8. 16 prioritized interrupts
9. Programmed and DMA I/O capability
10. N-channel silicon-gate technology

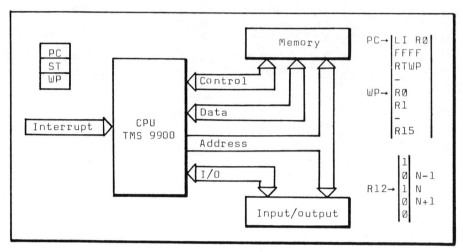

Fig. 3. The memory, I/O and interrupt function.

By exchanging the program counter, status register and workspace pointer, the microprocessor accomplishes a complete context switch with only three store cycles and three fetch cycles. After the context switch the workspace pointer contains the starting address of a new 16-word workspace in memory for use in the new routine. The old program counter, status register and workspace pointer are stored in this new workspace. This means that this microprocessor does not use a stack unlike most others.

Input/Output Bus

The microprocessor utilizes a versatile direct command-driven I/O interface, which provides 4k directly addressable input and output bits. Both input and output bits can be addressed individually or in fields of from 1 to 16 bits. The input/output bus employs three dedicated input/output pins and 12 bits of the address bus. The bus uses a bit serial synchronous technique.

PROCESSOR BOARD

The features of our processor board will be described in this chapter. The processor board is designed to hardware back-up the microprocessor TMS 9900 to be used in a multitasking system. Furthermore the processor board is designed to be used in the earlier described multiprocessor system, but due to the multiprocessor architecture this has only little effect on the processor board design. The multitasking capability of the processor board, on the other hand, has great influence on the processor board design. The multitasking hardware back-up is designed to make unwanted intertask interference impossible, and to improve memory relocation and protection.

Monitor/User Mode

In a multitasking system it is necessary to have an operating system. The operating system must have some privileges, so it can perform its job without interference from other tasks.

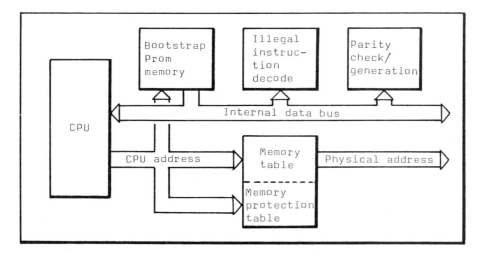

Fig. 4. TMS 9900 CPU board outline.

The operating system is said to run in monitor mode and all other tasks must run in user mode. In monitor mode the microprocessor and all peripheral equipment can be used to full extend. A task running in user mode is restricted to normal computations and must call the operating system to perform input/output, timing etc. If a task in user mode tries to execute beyond these limits, a context switch will result and the operating system will take control.

The processor board has hardware back-up for one task in monitor mode and seven tasks in user mode. These eight tasks are represented by a tasknumber (three bits wide) used for the memory table and memory protection table as explained later.

The tasknumber can only be changed in monitor mode. The processor mode is set to monitor mode on power-on. To change the processor mode to user mode it is necessary to run through a sequence of four instructions.

The processor mode will change from user mode to monitor mode if one of three conditions is satisfied:

1. A monitor function is called
2. An interrupt occurs
3. An error is discovered

When an error is discovered, an errorcode, the last instruction address and the last memory address is latched. This gives the operating system a possibility of tracing the error occured.

Physical/CPU Address

Using the processor board in a multi-processor system, it seems unreasonable to limit the processor memory space to 32 k words only. Expanding the number of address lines by only three, expands the processor memory space to 256 k words. The address coming out on the microprocessors address lines is called the CPU address and the address lines coming out of the memory address expansion unit is called the physical address.

We use a very fast RAM memory, called the memory table, to hold the actual relations between the CPU address and the physical address. Using the tasknumber as well as the most significant CPU address lines as input to the memory table gives a flexible memory management and a very fast context switch between different tasks.

The memory table can only be changed in monitor mode.

Memory Protection

The memory table described earlier is a pleasure to have but not a must. It is necessary to have some sort of memory protection to secure memory assigned to one task from being modified by another task. Due to the multiprocessor architecture and the memory table, we have placed the memory protection circuit on the CPU board in conjunction with the memory table. To each CPU address and to each tasknumber there is assigned a physical address and a memory protection code. If the memory protection code conflicts the actual memory

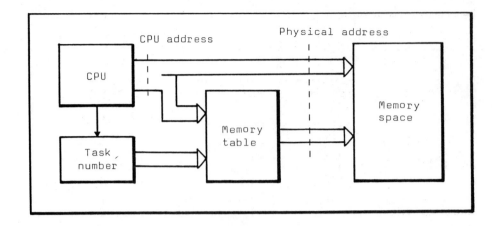

Fig. 5. Memory table usage.

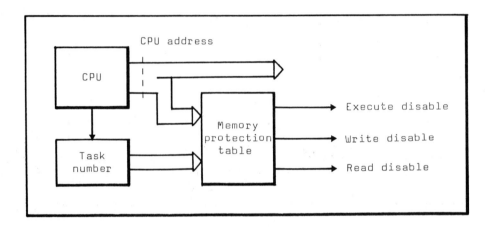

Fig. 6. Memory protection.

request from the CPU, the memory request is kept from reaching the bus and a context switch occurs, transfering control to the operating system.

The memory protection code can be set to one of four protection types:

1. INSTRUCTION AREA
2. DATA AREA
3. WRITE PROTECTED DATA AREA
4. ILLEGAL AREA

These four protection types give rise to three destinct disable lines (execute, write and read disable). The execute disable is active for protection codes 2, 3 and 4. The write disable is active for codes 1, 3 and 4 and the read disable is active for code 4. The memory protection will not allow instructions to be executed in an area where write is enabled. This will give a code protection as if the code was stored in a read only memory (ROM).

The memory protection table can only be changed in monitor mode.

Memory Address Errors

Because of the expanded addressing capability achieved through the memory management unit, the entire addressable memory is unlikely to be present. As a consequence non-existing memory detection by the CPU is implemented.

This function is obtained by allowing a certain period of elapsed time from CPU issue of memory request until ready acknowledge signal returns. When expired, a context switch is forced upon the CPU and a non-exist-

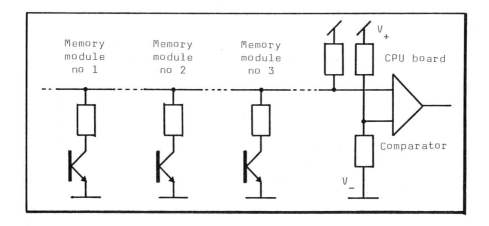

Fig. 7. Dual-existing memory test.

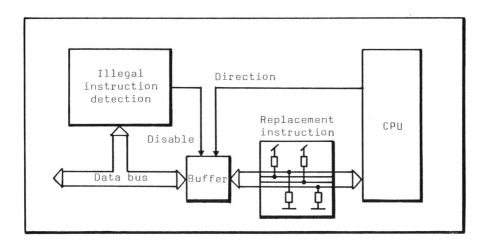

Fig. 8. Illegal instruction detection and replacement.

ing memory error code is latched.

Another problem can occur during the addressing of physical memory. Memory modules have a switch settable physical address. When two or more memory modules are used in a system, some modules may have identical address settings. This condition cannot be tested from the ready signal and an analog signal line has therefore been added. A memory module pulls this analog line down through a resistor, when addressed. The feed current is measured on the CPU board. If it exceeds a predetermined limit, this indicates that two or more memory modules have the same physical address.

The non/dual-existing memory detection is valuable during power-on, to determine the actual size of memory and to test that no two memory modules have the same physical address.

Memory Errors

In every microprocessor system there is a possibility for memory and transmission errors between the processor and memory modules. In our multiprocessor system there is an increased possibility for memory and transmission errors, due to the many buffers needed from the memory cell to the microprocessors data bus. To test both the memory and the transmission path between the memory cell and the microprocessor, each half memory word is expanded by one parity bit. The parity bits are generated and checked on the CPU board and stored on the memory board along with the data bytes. After a power-on, all memory cells have to be initialized in order to set the parity bits. If a parity error is detected on a memory read, the operating systems error handling routine is executed, at the end of the current instruction.

Illegal Instructions

Detection and replacement of illegal instructions is necessary to prevent a software error in one task from changing memory assigned to another task or from interfering with the operating system. The detection of illegal instructions is performed by using a field programmable logic array (FPLA). The instruction replacement is performed by not gating the illegal instruction to the microprocessors databus, in which case a resistor network defines the replacement instruction.

This logic array uses twelve of the data lines, the processor state (monitor/user mode) and an instruction aquisition line (IAQ - from the microprocessor) as input. On the output it presents a decoding of instructions needing special care and a strobe for illegal instructions. The processor state is used as input to allow a difference in monitor and user mode instruction decoding.
When an illegal instruction is detected, it is replaced by a monitor call, which automatically changes the processor state to monitor mode and starts execution of an exception handling routine.

In monitor mode the only illegal instructions are the undefined instructions. An undefined instruction is converted to no operation (NOP) by the CPU, but it indicates an error when happening.

In user mode there are several groups of illegal instructions, all of which would be fatal to the operating system if executed in user mode.

One group of illegal instructions concern interrupt mask setting. In user mode all interrupts must be enabled. Thus the "load interrupt mask immediate" (LIMI) instruction is illegal and will be replaced by a monitor call. The instruction "return workspace pointer" (RTWP) loads the interrupt mask. This instruction cannot be replaced, however, since it is needed in user mode as return from a subroutine. The actual interrupt mask loaded is therefore tested and in the event of an error a non-maskable interrupt will result.

All input/output must be controlled in monitor mode, making all input/output instructions illegal in user mode and replacing them if encountered in user mode.

The microprocessor has an instruction "execute" (X). The real instruction executed is fetched as data and is not tested as an instruction. On this background it is necessarily illegal in user mode as it by-passes memory protection and instruction error decoding. The instruction is replaced by a monitor call if encountered in user mode.

The last illegal instruction in user mode is the "branch and load workspace pointer" (BLWP). This instruction is used to call a subroutine. To enable the subroutine to set the needed interrupt mask, interrupt is not tested after a BLWP instruction. A user can make an infinite string of BLWP instructions, thereby disabling all interrupts. The BLWP instruction is needed in user mode and therefore we test for two consecutive BLWP instructions, and the second BLWP instruction in a sequence is illegal. This illegal instruction will be replaced by a monitor call.

Watch Dog

If the hardware, performing the earlier described error detection and handling functions, works perfectly, it should not be possible for an interrupt to stay unanswered in user mode during more than two instruction cycles. If this limit is exceeded a non-maskable interrupt is executed. This watch dog function provides redundance in the error checking hardware.

CONCLUSION

It has been shown that it is possible to make a safe process computer, that does not allow software errors from a user to interfere with the operating system or other tasks in memory. The error detecting and handling hardware is rather complex, but will only increase the software complexity slightly.

Novel microprocessors now appearing on the market are designed to operate in monitor/user mode and they utilize memory management. Privileged instruction facilities are also featured. Even multiprocessor environment is supported by a "test and set" instruction. Examples of such new microprocessors are the Zilog Z8000 and the Motorola MC68000. Using these microprocessors, much of the hardware complexity of this present CPU board design would be obsolete and similar functional performance would be achieved by far less effort.

Acknowledgment

The described work is carried out at the Laboratory of Automatic Control Systems, The Technical University of Denmark and supported by the Danish Council for Scientific and Technical Research, contract no. 516-8098.E-436.

REFERENCES

Caprani, O., K. H. Jensen, U. Ougaard (1977). Microprocessors connected to a common memory. Euromicro, 175-181.
Højberg, K. Søe (1977a). An asynchronous arbiter resolves resource allocation conflicts on a random priority basis. Computer Design, August, 120-123.
Højberg, K. Søe, J. R. Taylor and Chr. Suusgaard (1977b). Two-port microcomputer for a parallel processor. SIMS meeting, Studsvik, Sweden.
Højberg, K. Søe (1978). One-step programmable arbiters for multiprocessors. Computer Design, April, 154-158.
Lalive, Th (1978). TC8 Up-to-date report. European Workshop on Industrial Computer Systems. Obtainable from: Chairman, Dr. Th Lalive, Brown Boveri Research Center, CH-5405 Baden, Switzerland.
LeMalr, I., R. Nobls (1978). Complex systems are simple to design with the MC 68000 16-bit μP. Electronic Design 18, 100-107.
Simpson, W. D. and colleagues (1978). 9900 Family Systems Design and Data Book; first edition, Texas Instruments Incorporated, Houston, Texas.
Zilog Inc. (1978). Zilog Z8000 Advance Specification.

DISCUSSION:

Voges: I think your assumption that the operating system is correct, is dangerous, because in most cases the application software is easier to verify than the operating system.

Holt: Our application software is accepted non-tested, but the operating system is carefully tested and made after the recommodations given by Purdue Workshop TC-8.

Aitken: Did you check your memory protection against a random noised generator?

Holt: No, we have not done that. The protection is made in the way, that the computer can not access memory which it shouldn't access.

Aitken: So your memory is only protected by the operating system and not against influences from outside?

Holt: No, we have one parity bit for each memory byte and thereby influences from outside can be detected.

Parr: I want to expand a little bit on the statement of Mr. Voges concerning operating systems. The operating system is the heavest utilized part of the system and the smaller part of the program.

Voges: In most of the applications I am familiar with, the application programs consist of small modules and are mostly sequentiell, but operating systems are time dependent and are thefore hard to test completely.

Kirrmann: Why did you choose individual memories, addressable by any of the processors and not one general memory accessible by all processors combined with individual memories accessible by one only?

Holt: One way of using our memory protection in conjunction with the dual-port-memories and bus-switch, is to give each CPU a local memory and access to a pool of global memory. But our structure is not limited to this configuration.

Parr: In my opinion the solution for multi processor systems is to have local memories with access to one common memory, which isn't accessed very often.

Holt: The CPU, we have, has a lot of memory access every time, it executes an instruction. That's, why we have local memories.

Kirrmann: If you have a global variable, then you have it in every one of the local memories.

Holt: No, every CPU can access all memories through the Bus-switch.

CONTROL OF NUCLEAR REACTION BY PATTERN RECOGNITION METHODS

B. Dubuisson and P. Lavison

University of Compiegne, Division 'Control of Systems', BP 233, 60206 Compiegne Cedex, France

Abstract. The problem of surveillance of a complex system is a difficult one when security is also concerned. The work we present is a global approach of it. We use pattern recognition methodologies because these methods do not need necessary a mathematical model of the process. The performance of the control system was evaluated with experimental data from an High Flux Isotope Reactor from the Commissariat à l'Energie Atomique (France) for which this work was done.

Keywords. Signal processing ; pattern recognition ; nuclear reactor ; computer control ; man machine system

INTRODUCTION

The problem of surveillance of a nuclear reactor is not only concerned with the problem of keeping its power to a nominal value but also with security. So one have to be care with all data to monitor well all the system.

One can divide coarsly the surveillance problem into two parts :
- either a mathematical model is possible where all parameters are known or can be estimated. In this case a comparison between the observed data and the response of the model can determine if the observed state is correct or if there is a misfunctionning.
- either a mathematical model is not available (or the physical concepts are complex or estimation of parameters is difficult). So one must use all the information he can take out of the process. In this case, one use signal analysis, data processing and some others methods for decision theory, such as pattern recognition (Pau, 1974).

Our problem is concerned with an High Flux Isotope Reactor called RAPSODIE (Centre d'Etudes Nucléaires de Cadarache - France). There is a complex model of the running of the reactor, but this model depends of many parameters ; so it cannot be easily identifiable. So we propose to the responsible of this process an approach based on all the disponible information.

Many works have also be done on analysis of recorded signals : we propose a new methodology based on using pattern recognition processing (Gonzalez, Fry, Kryter, 1974) (Gonzalez, Howington, 1977). In a first part we describe this methodology, in a second part, the application done on the process and in a third part the results.

PATTERN RECOGNITION METHODOLOGY

Our problem is to test a process for normal or abnormal operation by observing all the possible data. The pattern recognition methodology can be decomposed into several steps (Figure 1).

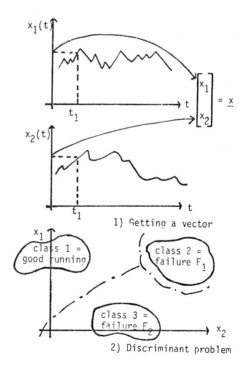

Fig. 1. Pattern recognition methodology

- getting a representative vector of the data
Several methods are possible : sampling the signals, using mathematical transforms (Fourier, ...), other descriptors. We call d the dimension of the vector.
- designing a discriminant function. Its aim is to classify vectors between several groups (called classes) : one group corresponds to the normal running of the system, the other ones to different misrunning. When each class is associated with a failure mode, a diagnostic of the system is possible. When there are two classes (dichotomy), one can only say : good running or bad running.
- testing this decision system by using well known samples and computing a criteria function.

For designing the discriminant function there are two possibilities :
- one knows all the statistics properties of vectors. So the Bayes decision rule can be implemented minimizing a cost rule of misclassifying.
If some parameters of statistics are not known, they can be suitably estimated from samples.
- the distribution law is not known, so one may use non-parametric methods. First one the Parzen estimate (Parzen, 1962) uses special functions called kernels to get a continuous estimate of the density function. Another and important one is the k-nearest-neighbor rule (Wagner, 1975) (Fukunaga, 1972) where an unknown sample is classified according to the class distribution of the k-nearest-neighbor samples. In this last case the techniques do not demand any knowledge about the distribution. However we have to pay for that advantage with more computation and the need of good samples for each class.

IMPLEMENTATION ON THE NUCLEAR REACTOR

The nominal power of the reactor is about 40 Mwth. This value can be increased or decreased by an action of the pilot. Many signals can be recorded on the process : input sodium temperature, 64 output sodium temperature, power, flow of sodium, ... For this first work we do on this plant, we choose to learn the power signal which is derived from a neutron flux sensor. By action on some specific commands, the pilot can adjust the power to any specific value.

First problem it is not possible to get the probability density function of any vector extracted from the signal. The Gaussian hypothesis can be formulated but no one is certain of its ability. So it is necessary to use non parametric methods.

These methods demand well-known samples that is a good definition of classes and many samples from each class. Good definition of classes is easy : we define a class of 'good running' and several classes of 'bad running', each one associated with a failure. More difficult is the problem of getting samples of each class except the first one. We must have samples of signals when the reactor is failing. Because of security it is impossible. This problem is well known : because of security or continuous running of the system, simulation of particular type of working is often not possible. So it is necessary to modify a problem formulation. A new approach is proposed on the assumption that the running of the system is explained by the plant's history. We know that the pilot can increase or decrease power plant by action on a specific command. So we define two classes in the power signal :
- signal corresponding to no action of the pilot (class NA) when the reactor is running at its nominal power.
- signal corresponding to an action (class A) when the power has been decreased or increased.

Samples of these two classes can be easily got. By observing the signal we class it into one of each class and give the information to the pilot. If the chosen class is NA and if the pilot have done some command or the opposite case, we say that the plant is in a wrong running. The scheme of this system is summarized in Figure 2 : it is a computer aided decision system including man.

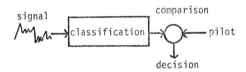

Fig. 2. Scheme of the decision system

As we said before there are three steps in implementing a pattern recognition decision system.

We have first to build a vector from the signal. We choose a very simple method that is to sample the signal during T seconds without doing any transformation (Figure 3).

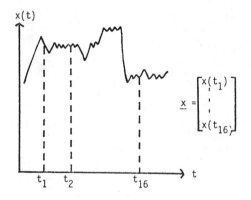

Fig. 3. Getting a vector from the signal

Dimension of the vector has been fixed by experience to d = 16 (time duration is 32 seconds). We build vectors (Figure 4) in the power signal corresponding to class NA and to class A.

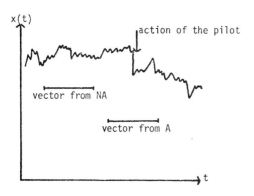

Fig. 4. Different vectors from different classes

Second step is to elaborate a classifier : we choose non parametric methods because no one knows the exact distribution probability of these vectors. The first method we try is based on the Bayes decision rule when using the Parzen estimator for the distribution probability.

Everyone knows the Bayes decsion rule when we use the (0,1) cost rule. We associate vector $\underline{x}$ to the class w_i where its a posteriori density $P(w_i/\underline{x})$ is maximum (Fukunaga, 1972)

$$\underline{x} \in w_i \text{ if } P(w_i/\underline{x}) > P(w_j/\underline{x}) \quad j \neq i$$

By using Bayes theorem :

$$P(w_i/\underline{x}) = \frac{P(w_i) \, P(\underline{x}/w_i)}{\sum_{i=1}^{M} P(w_i) \, P(\underline{x}/w_i)}$$

M is the number of classes.

$P(w_i)$ can be estimated by n_i/n (n_i : number of samples from class w_i ; n : total number of samples).

$P(\underline{x}/w_i)$ is estimated using the Parzen estimator by $\hat{P}(\underline{x}/w_i)$ (Parzen, 1962)

$$\hat{P}(\underline{x}/w_k) = \frac{1}{n_i} \sum_{j=1}^{n_i} \prod_{i=1}^{d} \frac{1}{h_1(n_i)} K\left(\frac{x_1 - x_1^j}{h_1(n_i)}\right)$$

x_1 is the l-th component of vector $\underline{x}$

x_1^j is the l-th component of sample vector x_j from w_i (n_i samples)

d is the space size

$h_1(n)$ are smoothing factor.

We choose $(h_1(n_i) = h(n))$

K(.) is the kernel function which is an arbitrary symmetrical bounded probability density function.

Under some conditions this estimator is asymptotically unbiased and consistent. A variety of kernel function or smoothing factors can be chosen. For instance we can search the best set which minimizes an error function :

$$e = E((P(\underline{x}/w_i) - \hat{P}(\underline{x}/w_i))^2)$$

We call it the optimum kernel function and optimal weighting factor. But the optimal weighting factor is dependent of P(..) through an integral. So one must use an interactive procedure proposed by Woodroofe (1970). Instead of using this complex method one can use the weighting factor proposed by Fukunaga (1972).

$$h(n_i) = n_i^{-k/d}$$

with $0 < k < 1/2$

The second non parametric method we propose to classify is the k-nearest neighbor rule (K-NN) : x is said to belong to class w : if the majority of its k-nearest neighbour belongs to w : this rule has good asymptotic properties.

The last step of this procedure is to check the method. We choose as a criteria function the probability of error that is the probability of misclassification. This function can be estimated by using a well known classified sample and testing it.

RESULTS

This method has been implemented on a minicomputer PDP 11/45 from Digital Equipment with 128K words of two bytes of core memory, two 2.5M words and two 40M words disks, an analog digital convertor, a videographic display unit and a graphic line printer. This complete system is necessary for analysis but not in exploitation. We give here values from the estimated probability of misclassification between class NA and class A. First when we use the Bayes rule with the Parzen estimate :

Kernel	1	2	3
Fukunaga h	0.40	0.40	0.16
Optimal h	0.40	0.395	0.156

$$1 : K(y) = \frac{3}{4\sqrt{5}} \left(1 - \left(\frac{y}{\sqrt{5}}\right)^2\right) \quad (y) < \sqrt{5}$$
$$= 0 \quad (y) > \sqrt{5}$$

$$2 : K(y) = \frac{1}{\sqrt{2\pi}} e^{-y^2/2}$$

S.C.C.S.—F

$$3 : K(y) = \frac{\sqrt{2}}{2} \exp{-2|y|}$$

Best results are obtained with the exponential kernel, the value of the smoothing favor does not seem important.

We try after the k-NN rule : the results are given for different values of k :

k	1	2	3
Probability Error	0.10	0.094	0.095

The probability of error does not decrease as k is more than three which seems the optimum value.

We implement the classifier in real time on the reactor giving the information to the pilot by a display. For this the power signal is sampled continuously to get sequential vectors which are classified either into NA either into A.

The system has been implemented for RAPSODIE and the probability of error evaluated for the 3.NN rule or for the Parzen estimator. Modifications of 0.5% of the nominal power can be easily deserved : this method permits to improve the problem of monitory the nuclear reactor and thereby to improve the security problem.

CONCLUSION

This work exhibits the interest of using pattern recognition methodologies in monitoring complex systems when security is an important problem. With some standards methods good results have been obtained in the surveillance of the power of a nuclear reactor. Developments are now possible such as using more than one signal or direct information from the pilot. The aim is to offer to the man a computer aided system which is interactive with him.

ACKNOWLEDGEMENTS

The author would like to thank Mr J. GOURDON from the Commisariat à l'Energie Atomique (Cadarache - France) from his helpful support.

REFERENCES

Fukunaga, K. (1972). Introduction to statistical pattern recognition. Academic Press

Gonzales, R.C., Fry, D.N., and R.C. Kryter (1974). Results in the application of pattern recognition methods to nuclear component surveillance. IEEE Trans. Nucl. Sci., 21, 750-756.

Gonzales, R.C., and L.C. Howington (1977). Machine recognition of abnormal behavior in nuclear reactor. IEEE Trans. Syst., Man & Cybern., 10, 717-728.

Parzen, E. (1962). On estimation of a probability density function and mode . Ann. Math. Stat., 33, 1065-1076.

Pau, L.F. (1974). Diagnosis of equipment failure by pattern recognition. IEEE Trans. Reliability, 3, 202-208.

Wagner, T.J. (1975). Non parametric estimates of probability densities. IEEE Trans. Inf. Theory, 4, 438-440.

Woodroofe, M. (1970). On choosing a delta sequence. Ann. Math. Stat., 41, 1665-1671.

DISCUSSION

Aitken : "Are your 4 signals all power signals ?"

Authors : "No, they are power, temperature, etc, ..."

Aitken : "So you are getting in a cross-correlation-situation. As I understood your presentation, you did your work, when the reactor was running in a steady state. Are you trying to do it under fault conditions ?"

Authors : "It is difficult to do that under fault conditions"

Genser : "Is your method for investigating the behaviour of the reactor only or is it for investigating the behaviour of the control system too ?"

Authors : "It is investigation of monitoring only"

Genser : "You showed your method for 4 signals. Have you investigated the problem of many signals ?"

Authors : "I could only investigate this simple configuration. It is rather difficult to get more data".

SPECIFICATION, DESIGN AND IMPLEMENTATION OF COMPUTER-BASED REACTOR SAFETY SYTEMS

M. J. Cooper, W. J. Quirk* and D. W. Clark**

Materials Physics Division, AERE, Harwell, Didcot, Oxfordshire, England
*Computer Science & Systems Division, AERE, Harwell, Didcot, Oxfordshire, England
**Department of Engineering Sciences, Oxford University, Oxford, England

Abstract. Computer-based reactor safety systems must be fault-tolerant in order to provide high levels of availability and safety. However, since such systems must respond very quickly and reliably to potentially dangerous situations there are certain constraints which limit the way in which the required fault-tolerance can be achieved. These constraints are considered and techniques which can be used in the specification, design and implementation of such systems are discussed.

The software for a reactor safety system can be partitioned into a number of programs which interact only through the transfer of data from one to another. Such a system can therefore be implemented in a number of ways, for example by partitioning the hardware in the same way using microprocessors, with a separate processor for each program. We have therefore developed a theory for analysing the reliability of such a system, in order to assess and compare different possible architectures, and a format mathematical framework which provides the facility to prove, in a mathematical sense, the integrity of a proposed implementation.

A self-checking dual processor and an efficient coding technique for fault-tolerant data transfer have also been developed for use in high reliability systems.

Keywords. Reactor Safety Systems; Fault-tolerant Computing; Computerised Safety Systems.

INTRODUCTION

A reactor safety system is required to monitor the output from a large number of sensors and to interpret the sensor readings in terms of a safe or dangerous condition. Although there may be more than one type of sensor, for example temperature and flow rate sensors, there may well be several thousand sensors of a given type, with the data processing for each of these being essentially identical apart from differences in the numerical value of certain parameters. Moreover, the processing requirements in deriving a safe or dangerous condition from a set of sensor readings can be broken down into a number of separate processes which are in principle fairly simple. The data analysis function of the safety system is thus ideally suited for implementation using computers, since the simple algorithms can be applied for each sensor reading without replication of hardware to the extent necessary in a system which provides a hard-wired unit for each sensor.

A reactor safety system must be highly reliable in two respects: it must have high availability so that the probability of an unscheduled reactor shutdown due to failure of the system is small and it must have a high degree of safety so that the probability of an accident occurring as a result of an undetected dangerous situation is extremely small. The maximum permissible values for these probabilities may be of the order of $0.05/yr.$ and $10^{-10}/yr.$

respectively, the latter corresponding to a mean fractional dead-time of about 10 . The achievement of such reliability requires a high level of fault-tolerance in the system and this must be considered carefully in the specification, design and implementation of the whole system, including both the hardware and the software. Moreover, since the system must respond to the detection of a potentially dangerous situation by shutting down the reactor within a very short period, it is vital that achievement of the required fault-tolerance does not degrade this response.

FAULT-TOLERANCE

In general fault-tolerance is achieved by the provision of some form of protective redundancy in the system which enables errors to be corrected or masked. This redundancy may therefore involve redundancy in the components of the system, either hardware or software, or redundancy in time. Component redundancy may consist of replication of units to give multiple outputs from identical operations which can then be compared for consistency. An example of this technique is the use of triple modular redundancy (TMR) with majority voting on the outputs. This provides a masking of individual errors and an indication of which component is faulty. However, it must be emphasized that this technique is vulnerable to common mode failures, particularly if all the units are identical.

In order to reduce the likelihood of common mode failures multiple version redundancy can be used, where each component uses a different method for determining the output value. An example of this is the use of n-version programming (Chen, 1978). Since many of the software errors may arise in the programming stage the use of different programmers, algorithms or languages can reduce the probability of multiple erroneous outputs. Such redundant software modules may be used in parallel, with comparison of their outputs, or in series, with alternative versions used only if errors are detected in the output provided by the version in normal use. However, this technique will not eliminate common mode failures due to errors in the original specification.

A different approach to fault detection is the addition of extra data or software. Redundant data may involve the use of, for example, parity, Hamming codes or cyclic redundancy codes to provide acceptability checks on data. Redundant software can provide for routine testing of the correct operation of the system by either introducing special test programs or carrying out normal operations on special test data.

Redundancy in time, in which a process is repeated if the output obtained from it is erroneous, relies on the detection of errors in the output from a single process and the ability to return to a previous error-free state. It may involve repeating the same process with identical input data or the use of in-series redundant software. In either case a time overhead is introduced which is dependent on the rate of occurrence of errors in the system.

The above techniques, several of which may be used together, can provide error masking, thus giving tolerance of transient faults. However, tolerance of permanent faults requires in addition that a permanently failed component is replaced, either by direct replacement or system configuration, unless sufficient redundancy is present to ensure an adequate level of multiple fault-tolerance.

REACTOR SAFETY SYSTEMS

The over-riding requirement for a reactor safety system is that it should reliably detect a potentially dangerous situation and should respond within a very short period so that the reactor can be shut-down before an accident can occur. The most serious constraint on the achievement of the required fault-tolerance is therefore one of time, since it is essential that the response is not subject to uncertain delays. This means that any techniques used to obtain fault-tolerance must be consistent with well determined timing constraints. Thus, for example, it is not possible to employ roll-back techniques which are introduced only when errors occur and may add an uncertain delay to the completion of the safety function, particularly if a number of transient faults can occur during a single monitoring cycle.

The other main constraint on the achievement of fault-tolerance in a reactor safety system concerns the replaceability of failed components. Since the required levels of availability and safety are very high it is unlikely that any dynamical reconfiguration of the system during normal operation would be permissible. Replacement of failed hardware components may be possible, but this would probably be restricted to direct replacement by an identical component to take over the same function. Furthermore, some components of the system, for example the temperature sensors, will be inaccessible during normal operation, so that the system must be capable of functioning correctly with a certain number of its components in a failed state.

Component redundancy is thus essential for two reasons. Since the sensors are inaccessible each independent sensor must be replicated to such an extent that all expected sensor failures can be tolerated until routine shut-down periods. Replication of components of the processing system is also necessary so that errors can be masked and the main safety function is not impaired by the introduction of delays into the processing or by lack of physical resources.

SYSTEM ARCHITECTURE

We have shown (Cooper, 1978) that the basic function of the system can be subdivided into a number of separate functions and that the software can be partitioned in such a way that individual programs need communicate with each other only through the access by one program of data which have been stored previously by another program. It is therefore possible to partition the hardware in a similar way and, since each program in itself is fairly simple, it is feasible to implement the system using microprocessors with each basic program operating in a different processor and each input processor servicing up to perhaps 100 sensors. Such an architecture therefore eliminates the need for an overall operating system as such, thereby avoiding one common source of errors. It also enables identical modules to be used for different components, thus providing ease of replacement of failed components, which should also increase the reliability of the system.

If the readings from a large number of sensors are analysed by a system comprising a fairly large number of processors there is clearly a number of ways in which the system can be configured. Even with the processing partitioned in a given way there is still a number of ways of distributing the input data between the various input processors. In order to be able to quantify the reliability considerations and to compare various possible configurations we have therefore developed a theory for analysing the availability and safety of such a system based on the reliability of individual components (Cooper, 1977). We have assumed that it will be impossible to replace or repair sensors whilst the reactor is operating but that failed processors can be replaced within a short period of their failure and we have therefore extended conventional reliability theory (Basovsky, 1961) for the case of a high reliability system containing both replaceable and non-replaceable components.

As a specific example we have considered a system in which the software can be partitioned into three separate programs, as illustrated in figure 1:

Program 1. Reads the sensor level, checks it against current limits and sets an alarm flag appropriately, adjusts the limits if required, organizes sensor tests and sets a sensor status flag.

Program 2. Combines the alarm and status flags for the sensors in a given group and determines a safe or dangerous condition.

Program 3. Carries out long-term checks on the limits and determines whether these indicate a dangerous condition.

Each program can then operate in a separate block of redundant processors, with sufficient blocks to analyse data from all sensors. Initially we have considered the two limiting input configurations such that inputs from each sensor in a given redundant group are connected to the same processor (scheme a) or to different processors (scheme b), as illustrated in figure 2. We have not at this stage included any multiplexing between the sensors and the processors and for convenience we have considered memories to be effectively part of the processors which have write access to them.

The theory provides expressions for the reliability functions of the system, such as unavailability and accident probability as a function of the component reliability parameters and the redundancy factors. In order to compare the different architectures quantitatively we have evaluated the reliability functions for example values of the parameters. The results indicate that, for the range of reliability required, scheme (b) is always superior to scheme (a), since the latter is more dependent on the processor reliability. This is consistent with the expectation that the level of fault-tolerance will be higher if inputs from redundant sensors are segregated through different channels in the processing system. In both cases the most important factor is the reliability of the sensors, since these cannot be replaced, and the redundancy factor for these will need to be sufficiently high for the system to be able to tolerate the number of sensor failures which will be expected to occur during a single operating period.

This analysis thus indicates that the major limitations on the system reliability arise from failure of the sensors and the way in which they are connected to the processing system. It also supports the belief that sufficient tolerance of hardware failures can be achieved through a reasonable level of component redundancy when the system is partitioned into a number of simple programs operating in separate processors.

SPECK
A SYSTEM SPECIFICATION AID

While much weight is rightly placed on a detailed examination of how hardware failures will affect a safety system, it is also clear that the software must achieve a very high standard of correctness. We have been concerned with several aspects of this problem area. Firstly there is a need for a vehicle by which engineers and system designers can communicate in an unambiguous manner. Secondly, certain types of inconsistency in a specification can affect fundamentally any system which meets this specification in a way which seriously jeopardises the integrity of that system. Thirdly, analysing the effect of changes in the behaviour of subsystems, whether brought about by hardware failure or software inadequacy, provides an approach to attaining the high degree of confidence required in the total system. Fourthly, the proof or demonstration of any system demands a complete specification which is the yardstick by which the system is evaluated.

The SPECK approach to these problems is founded on a functional model of the system (Quirk & Gilbert, 1977; Quirk, 1978). The functional requirements of the system are synthesised from a network of simpler subsystems or processes. The result is an abstract implementation of the system and because of the way these abstract processes interact, the resultant abstract system forms a mathematical model of any implementation which meets this specification. The power of this view of the specification is that, like any mathematical model of a system, properties of the system are predicted - or at least predictable - from the model.

The SPECK model can be seen as the most abstract possible model of the system in the following sense. Because of the nature of the SPECK processes and because the description of these processes incorporates realistic timing properties, one can visualise a real system obtained by implementing each SPECK abstract process as a real process - real software running on real hardware. On the other hand, none of the properties of the individual hardware or software units are assumed by the model, other than that they allow the timing constraints put on the processes to be realised. As a result, the properties of the real system reflected in the SPECK model are only those which follow from the topology of the system - that is, the possible interaction paths between the processes and the timing interrelationships forced upon them. This ability to handle timing constraints in a straightforward manner is a powerful attribute of the method and the mathematics has been developed to answer quite categorically whether or not such a specification is self-consistent. The implication of this is that since the real-time behaviour of the system is predicted by the model, the implementation of a system which meets a self-consistent SPECK specification has a guaranteed and provable performance, at least in the absence of failures.

The verification of model consistency and indeed the production of the model itself can be quite tedious. However, the mathematics involved in the proof of consistency is constructive and so an automatic analyser for the SPECK language has been constructed. This runs on a 16 bit minicomputer and though only a prototype, it has demonstrated quite clearly that computer aid is feasible in this area. The analyser is an adaptive, interactive tool which will guide a user through the specification procedure, building up the system model in the course of this interaction.

A specification for a system usually - and hopefully - contains not only the complete functional requirements of the system but also various safety, economic and other requirements which are properly constraints on the design of a system to meet the functional requirements. No one can produce a system which will not fail, but given the failure rates of the various physical building blocks of the system, one can construct a system of arbitrarily low (but non-zero) failure probability. If one considers a simple TMR system, a single unit is replaced by three similar units, and while a single failure puts out the single subsystem, two failures are needed to put out the triple system. From a functional point of view, the TMR system can be seen as a unit within which the implementation of any function for one of the three subsystems still leaves the overall transfer function correct. By using the SPECK technique, one can investigate the ways in which the functional behaviour of the overall system can be made independent of certain individual functions within the system. This analysis is carried out without any assumptions about the way such processes fail. As abstract processes do not fail, we model real failures by allowing a them to arbirarily modify the function calculated by that process. Since the processes in the model do have to be mapped eventually on to real processors, this approach allows one to decide to what extent the processes need to be divided among different processors to achieve acceptable reliability. Notice that this independence of total system function is maintained through software inadequacy as well as hardware failure.

Finally, the ability of a system to respond to given transient events and within a given time can be determined from the model. Overall sysem response is a function not just of the topology of the system but also of the functions implemented by the individual processes. The computer representation of the model produced by the SPECK analyser allows a complete simulation of the system which yields the real-time behaviour of the system when it is implemented with a given set of algorithms. Thus not only is there a test-bed facility for developing and checking such algorithms, but again a categorical answer can be given to the question of the system's response to given external stimuli.

Thus we believe that the SPECK approach is the first in a line of automated tools which will enable a system to be designed which has provable properties visible in and implied by a provably complete and consistent specification.

SELF-CHECKING PROCESSORS

A system which functions correctly in the absence of faults must also tolerate a sufficient number of hardware failures. In order to reduce the level of redundancy necessary to achieve this, detection of a fault and identification of the failed component are necessary in addition to tolerance of the fault, so that permanently failed components can be replaced quickly. We are therefore investigating the use of self-checking components in this type of system.

One possible method for detecting faults in a microcomputer module in the safety system is to cause it to execute a CPU check program after each cycle of data processing operations. Any detected error could cause the CPU to loop or halt, and a watch-dog timer would then flag the failure to the rest of the system. The use of software in this way relies implicitly on comparing the results of one group of instructions with the results of another group. Before the advent of LSI, CPU checking programs were based on an intimate knowledge of the circuit layouts and data paths within the processor, and it was possible to test each gate through all input combinations assuming that all other gates were functioning correctly. With a microprocessor, however, a faulty gate is liable to affect the operation of the circuitry in its vicinity, so it is possible to imagine failure modes which are undetectable by even the most extensive code. Most importantly, it is essentially impossible to quantify exactly the probability of failing to detect a CPU fault. This is unimportant for many applications, where even the simplest CPU checks are worthwhile, but it is crucial in a reactor safety system.

Testing of complex logical circuits is often accomplished by supplying a representative sample of different inputs to the circuit under test, and to a circuit which has been shown to work correctly previously. The outputs of the two circuits are compared, and allowing for deskewing, any discrepancy between the outputs indicates an error in either one or other of the circuits. An error signal may also be produced by a fault in the comparator or error indication circuitry. The outline of an implementation of this technique to self-checking microcomputers (Cope, 1979) is shown in figure 3. The hardware consists of two autonomous microcomputers together with a master clock, a comparator and a latch to record transient errors. The microcomputers are self-contained to reduce the possibility of common-mode errors which the comparator would be unable to detect.

The comparator could monitor all the address, data, and control buses, or more typically simply the data bus. It should itself be checked to prevent its failure from obscuring a more serious microcomputer failure. As the comparator should always give '0' outputs for identical inputs, a programmable inverter is inserted to complement one set of inputs periodically, to check for a 'stuck-at-0' condition. The signal should be asynchronous with the master clock, and could be taken from the watch-dog timer. The remaining circuitry checks that the computer outputs are all 1's or 0's in accordance with the control signal to the inverter, and discrepancies are clocked into a latch by a signal derived from the microcomputer's clock. A further latch is used to maintain an error indication. As the circuitry following the comparator cannot be tested through all possible states it should be replicated for reliability.

The self-checking microcomputer pair is the standard building block of a possible hardware implementation of the safety system. Any error of a single microcomputer is immediately detected and flagged so that maintenance procedures can be initiated. One possibility is to allow a human operator to initiate a retry operation to see whether the fault is transient or permanent, and then to replace the complete unit in the latter case. The redundancy in the rest of the safety system means that it is still functional even if some microcomputer pairs are out of action.

The self-checking pair essentially trades off availability against improved safety, so it is important to discover the frequency of transient errors which are recoverable in a practical system. An informal attempt was made to do this, in which a pair based in Intel's SBC 80/10 microcomputers was combined with a comparator and (four-way) replicated detector. The microcomputers were programmed to run a continuous CPU check which exercised most of the 8080 instructions. (It is interesting to note that the CPU check took 650 bytes, whereas the functional programs for the safety system are typically 100-200 bytes long).

No special precautions, such as cooling, a special power-supply or shielding, were taken, and the system was kept running for some months in an office environment. The mean-time between transient failures (thought to be mainly due to mains fluctuations) was of the order of several weeks and recovery occurred on retry on each occasion. Whilst this statistic has no particular value, it lends some weight to the view that the mean-time between transient errors could be made acceptable using elementary design precautions in a final version.

DATA COMMUNICATION

When partitioning the safety system into a set of microcomputers, it is important to ensure that the communication of data be at least as reliable as the other components of the system. This of course could be achieved using sufficiently complex coding, but the question arises as to whether hardware or software encoding and decoding methods should be used. In the end, software coding was chosen as it could be achieved reliably within the self-checking pairs, whereas hardware coding would itself have failure modes which could be undetectable (Maghsoodi, 1979).

The data to be communicated consists of sets of byte-pairs, as 16 bits are typically required for representing temperature at a given site. Both the data and a 16 bit address are encoded as a block, and both convolutional and block-parity codes were considered. The block-parity code proved to be particularly appropriate, as encoding and decoding takes about 40us only (on a Z80 microprocessor) and imposes only a modest overhead to the main algorithms. The encoding of the address with the data means that 'burst' errors due to faulty address lines to the memory can be detected as well as faults in the memory itself. The block coding provides some degree of error-correction, giving a modest improvement in availability, but its main role is to enhance error detection, which for the specific coding used can be many orders of magnitude better.

CONCLUSIONS

The techniques adopted to achieve the required level of fault-tolerance in computer-based reactor safety systems must be such that response to potentially dangerous situations cannot be adversely affected by the occurrence of individual faults. In particular this precludes the use of roll-back techniques which may add uncertain delays into the processing period. Acceptable techniques must therefore involve redundancy in components without redundancy in time, unless the latter is constant. Since the sensors cannot be replaced whilst the reactor is operating, the level of sensor redundancy must be sufficient to tolerate all expected sensor failures until routine shut-down periods. On the other hand, it is possible that the components of the processing system can be replaced whilst the reactor is operating and we have therefore extended conventional reliability theory to enable the reliability of the system as a whole to be evaluated for different possible architectures.

The reliability of software for a computerised reactor safety system will depend on both the correctness of the programs and their tolerance to hardware failure in the system. The functional requirements of the system are such that the software can be partitioned into a number of simple programs which can be proved error-free with respect to their individual specifications. However, the overall system involves complex real-time interactions between the programs and there is therefore a need for proof of correctness of the specification as a whole, since this may easily contain logical design faults. A formal mathematical framework has therefore been developed to allow such specifications to be defined unambiguously and tested for consistency and completeness.

In order to provide fault detection in a redundant processor system a self-checking dual microprocessor has been designed and implemented in a unit based on the Intel SBC 80/10 microprocessor. Methods of ensuring reliability of data transfer between processors have also been considered and an efficient software coding method has been developed using a Z80 microprocessor.

ACKNOWLEDGEMENTS

The work discussed in this paper forms part of a project to carry out studies related to highly reliable computer-based control and protection systems. Contributions to this work have been made by a number of our colleagues at Harwell and Oxford, in particular M. Cheeseman, R. Gilbert, C. J. Kenward, A. Langsford, S. B. Wright, S. N. Cope and R. Maghsoodi.

REFERENCES

Chen, L. and Avizienis, A. N-version programming: A fault-tolerance approach to reliability of software operation. Eighth International Conference on Fault-Tolerant Computing. Toulouse, France (June 1978).

Cooper, M.J. and Wright, S.B. Some reliability aspects of the design of computer based reactor safety systems. IAEA / NPPC Specialists' Meeting, Pittsburgh, July 1977. AERE-R8927.

Cooper, M.J. and Cheeseman, M. Reliability and architecture of plant safety systems. AERE-R8959 (1977).

Bazovsky, I. Reliability Theory and Practice. Prentice Hall Inc. (1961).

Quirk, W.J. and Gilbert, R. The formal specification of the requirements of complex real-time systems. AERE-R8602 (1977).

Quirk, W.J. The automatic analysis of real-time system specifications. AERE-R9046 (1978).

Cope, S.N. A microcomputer-based reactor safety system. OUEL Report No. 1159 (1976).

Maghsoodi, R. Software coding for reliable data communication in a reactor safety system. OUEL Report No. 1264 (1978).

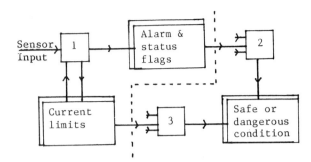

Fig. 1 Partitioning scheme for analysis of data from a single sensor.

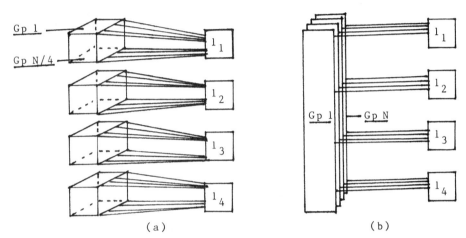

Fig. 2 Input configurations for schemes (a) and (b).

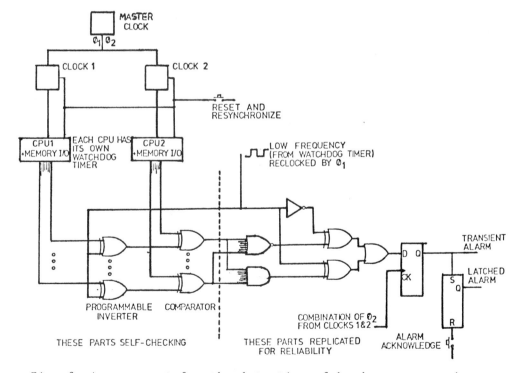

Fig. 3 Arrangement for the detection of hardware errors by comparison of data and output highways of two microcomputers

DISCUSSION

Trauboth: Have you actually implemented the system on computers in order to make some experiments and to demonstrate the effectiveness of such a system?

Quirk: We haven't got the whole system implemented at present and we aren't really allowed to play with reactors. There is a fairly complex "test-rig" to be build in next year or so and this system may be put on there. But at present we have no particular experiences with the specific system shown here and this is only an example.

Seidler: Do you check the CPU only for go/nogo or do you have some diagnostics? Do you also test the memory?

Quirk: We indicate only the go/nogo situation of the processors at present. The memory is not checked.

Lauber: How did you optimize the number of six sensors, why not five or seven?

Quirk: We have an estimate of the safe and dangerous failure rates of the sensor. To achieve a given reliability, this fixes the minimum number of sensors in each group which must still be operational at the end of the reactor cycle. The initial number in the group is then determined by the availability target to be met. With our figures, 6 was the least number of sensors to achieve both targets.

Göhner: Speck is a specification language to describe complex safety-systems. Are there any means to develop such a specification?

Quirk: We have certainly found in our experience that a 'top-down' decomposition into subsystems is the most natural way to proceed. Enhancements to the speck system will promote this modularity, but we cannot offer a 'methodology' at present.

EXPERIENCE WITH A SPECIFICATION LANGUAGE IN THE DUAL DEVELOPMENT OF SAFETY SYSTEM SOFTWARE

H. So, C. Nam, H. Reeves*, T. Albert**, E. Straker**,
S. Saib*** and A. B. Long****

University of California at Berkeley
*Babcock & Wilcox
**Sciences Application, Inc.
***General Research Corporation
****Electric Power Research Institute

INTRODUCTION

The validation of computer software is a complex and costly process which is becoming increasingly important to the nuclear industry as the dependance on software grows. The recent discovery of programming errors in two seismic design codes, which led to the shut down of a number of U.S. nuclear plants, is symptomatic of the magnitude of the problem.

In late 1977, the Electric Power Research Institute initiated research project 961, "Validation of Real-Time Software for Safety Applications". The University of California at Berkeley, Babcock & Wilcox, Science Applications Inc., and General Research Corporation are the contractors. This project is focusing on developing and testing methods for quantitativeley assessing the correctness of software to be used in real-time safety system applications. Specifically the objectives are:

1) To formulate a methodology which enhances the correctness and testability of such software.

2) To develop the necessary tools to support this methodology and quantitativeley assess correctness.

3) To demonstrate the methodology and tools on a representative application.

4) To evaluate the cost-effectiveness of this approach relative to more conventional techniques.

This paper presents some of our preliminary experiences with using a formal specification language in the dual development of preliminary designs for a safety application.

REPRESENTATIVE APPLICATION

Fig. 1 is a simplified schematic diagramm of a portion of a Pressurized Water Reactor safety system instrumentation. In general, quadredundant channels each with independent sensors are used to monitor specific plant parameters. Each channel contains the appropriate logic to compare measured or derived parameters against predefined limits and independently generate a trip signal when limits are exceeded. The four sets of trip signals are then fanned out through isolators to four sets of redundant voters which will initiate protective actions on the basis of two-out-of-four logic.

The safety system software application selected for this project is a calibrated neutron flux signal device. It provides a continuous calibration of the excore neutron detectors accounting for rod position, moderator density and other factors by using the more accurate but slower core thermal power measurements. The application will be implemented with 1000-2000 lines of Fortran code.

DEVELOPMENT METHODOLOGY

A flow diagram of the software development and validation methodology which was formulated and is being investigated in this project is shown in Fig. 2. The major features include:

1) The use of two independent teams to develop preliminary designs and detailed coding from the same functional requirements.

2) The use of a third independent team to coordinate the activities and to test and intercompare the intermediate an final results without contamination.

3) Integrated testing will be performed by all three teams.

4) A quantitative assessment of correctness will be made using representative testing and symbolic execution.

5) A variety of tools are being used to support this approach:

- Specification Language for the preliminary design.
- Automated test case generator and dual comparator for representative testing.
- Symbolic executor.

The methodology has been divided in more detail elsewhere. Currently, the dual preliminary design reviews are being completed and the coding is starting. The various tools have been finished and tested. This paper will discuss the initial experiences the contractors have had in the functional requirements and preliminary design phases of the project.

FUNCTIONAL REQUIREMENT PHASE

The requirements document is a structured document written in free form English by the Nuclear applications specialist. It is frequently not possible in these complex applications to develop the requirements in closed form. Consequently, the engineer must state them as precisely and unambiguously as possible and then thoroughly validate the concept through plant simulations.

The requirements were formulated in two major sections by Babcock & Wilcox and accompanied by nine sets of detailed acceptance test data:

Design Bases: set forth environment and the physics of the problem.

Specification of Requirements: represent a specific statement of the requirements and associated algorithms which have been thoroughly validated by the applications specialist.

The functional requirements for the Calibrated Neutron Flux Signal Device are expressed as 28 separate requirements and 16 more general rules or criteria. There have been 6 sets of major revisions since the first document was placed in configuration control over one year ago.

FUNCTIONAL REQUIREMENTS -

LESSONS LEARNED

More than one-half a man year has been spent to date modifying and revising the functional requirements document.

This document has gone through a number of significant evolutionary stages: first to add supplementary information required by the software engineer to understand the basis of the requirements, second to state explicitly the required algorithm and control logic, and third to separate design bases from actual functional requirements. As the project progressed it became clearer that two independent specification and programming teams required a precise and unambiguous statement of all requirements. Specifically:

- The algorithm had to be explicitly stated and validated by the applications specialist,

- Control logic was stated in decision tables in order that there would be no ambiguity about systems state.

All requirements should be testable, and for this purpose presumably correct acceptance data was provided to check the algorithms. However, successful execution of this data does not necessarily mean that the application is correctly coded.

The large number of revisions to the functional requirements document has emphasized the cost of configuration control and stressed the need for automated documentations control.

PRELIMINARY DESIGN PHASE

Two teams independently developed preliminary designs using the Requirements Specification Language (provided by the Ballistic Missile Defense organization). A third organization coordinated the process and analyzed both preliminary designs before detailed design or coding commenced. The objectives for this approach were four-fold:

1) preliminary design emphasizes systematic decomposition of the problem before detailed coding can start

2) dual preliminary designs provide redundancy

3) an independent third party permits independent comparison of designs and independent testing (without contamination)

4) the use of a formal specification language removes ambiguity and facilitates independent review.

The emphasis placed upon the preliminary design phase in this project is designed to detect software errors early in the software development cycle when they are easier and much less expensive to correct.

REQUIREMENTS SPECIFICATION LANGUAGE

The Requirements Specification Language (RSL) and the Requirements Evaluation and Validation System (REVS) developed by TRW for the Ballistic Missile Defense Advanced Technology Center were used in this project in order to avoid ambiguity, promote precision, and facilitate the review of the preliminary designs.

RSL is based upon four primitive concepts: elements, attributes, relations and structure. Examples of each are:

Elements = data, file, entity, Alpha (RSL terminology for submodule), etc.

Attributes of Data = initial values, units, etc.

Relationship between two elements = data input to Alpha, message passes through input interface, etc.

Structure can be attributed to an R-NET to define sequence of processing steps

Using this language it is possible to define: the functional capability of submodules (free text or by assertions), the control flow among modules, the definition and flow of data, and the traceability back to the originating functional requirements. The support tools provide automatic syntax checking an report generation. In addition, they assist in analyzing design structure and control and data flow.

USE OF RSL IN THE PROJECT

The use of RSL in this project will be dicussed from the perspective of the two dual development teams and from the perspective of the independent tester.

For the developer, RSL was found to provide the following assistance:

1) Traceability ensured that every RSL element related to a specific portion of the functional requirements, and that all requirements were decomposed to some RSL elements.

2) The R-NET structure provided in RSL ensures that the control logic interconnecting submodules is unambiguous and that all alternatives at each decision point are addressed.

3) The analyzer, report generator and sorting capability of RSL eliminates syntax errors and help to ensure the completeness and consistency of element attributes. As an example, when a specific functional requirement is modified, REVS can be used to list all RSL elements tracing to that functional requirements.

For the tester, RSL provides an effective audit toll; however, by itself it cannot establish the correctness of the preliminary design. Specifically, RSL was used to generate sorted reports for:

1) Intercomparing requirements and preliminary designs

2) Establishing that all required attributes had been provided for each types of RSL element

3) Walk-throughs and limited manual executions of the design with simulated input.

RSL does not eliminate the need for verification of the correctness of the preliminary design; however, it facilitates this review by eliminating ambiguity.

RSL PATH ENUMERATION

Fig. 3 presents the RSL path enumerations for the preliminary designs developed by both teams. As shown in the legend, paths preceeded by a solid dot are ORed and depending on the conditions one or the other will be executed. Paths intersecting without a preceeding dot are ANDed and must all be executed.

Although both preliminary design configurations are similar in layout, design 1 is decidedly more complex. This is because the Team 1 chose to explicitly represent certain logic in RSL which Team 2 decided was best placed within a number of separate submodules. Note, that the details within a submodule (alpha) are not known to RSL although they may be defined through the use of descriptions or input/output assertions. With the exception of this point, both teams otherwise included about the same level of control detail in the preliminary designs.

PRELIMINARY DESIGN STATISTICS

Certain RSL design statistics are summarized for both Teams 1 & 2 in table 1. Team 1's more complex control structure is again indicated by the larger number of paths. However, note that both teams decomposed the application into a similar number of submodules or alphas.

There were two other significant differences between the preliminary designs which are not evident in this data:

1) At considerable expenditure of manpower, Team 1 incorporated large portions of the functional requirement text within the preliminary design. Team 2 chose only to reference the functional requirements by paragraph number.

2) Team 2 used input/output assertions to define the function of each submodule (alpha). Neither team included much other description of the submodule functions.

The significant difference in manpower expenditure can be attributed to the level of detail in documenting functional requirements and data types and to the difference in the organizations. The errors which have been uncovered to date by the independent review team primarily involve misinterpretation of control and computational sequencing.

PRELIMINARY DESIGN EXPERIENCES - BENEFITS

The preliminary design phase provides for an orderly decomposition of the functional requirements. Attention is focused first on the global software structure including submodule definition, interface and sequencing; and on data definition and flow. From this the programmer can subsequently develop and test the more detailed code.

The use of RSL/REVS in this project provided an unambiguous framework for expressing the preliminary design. The formal structure of RSL imposes a disciplined design process, and the automated syntax checking and report generation capability of REVS reduces the likelihood of error and the amount of engineering effort. The RSL preliminary design was felt to be easier for the reviewer to understand and examine. Using the sorting and selective listing capabilities the tasks of auditing and review are simplified.

The resulting preliminary design document should also be easier to maintain again using the capabilities of REVS.

PRELIMINARY DESIGN STATISTICS

	TEAM 1	TEAM 2
RSL STRUCTURE		
N⁰ R-NETS	3	1
N⁰ SUB-MODULES (ALPHAS)	24	28
INTERFACES	2	2
PATHS	448	36
MANPOWER EFFORT		
LEARNING PERIOD	4 MW	3 MW
INITIAL DESIGN	30 MW	9 MW
MODIFICATION & CORRECTIONS	15 MW	2 MW
INDEPENDENT REVIEW EFFORT	3 MW	3 MW

PRELIMINARY DESIGN EXPERIENCE - PROBLEMS OR DISADVANTAGES

The lack of a clear differentation between preliminary and detailed design phases produced insecurity for the two design teams over the required level of detail. The major differences in the resulting preliminary designs involved the use of assertions, incorporation of functional requirements and definition of data.

The use of a formal preliminary design language requires a substantial commitment of manpower to learn about the system and to perform the necessary mechanics of inputting a large amount of design information. Although this procedure has the associated benefits discussed in the preceeding slide, the cost-effectiveness remains to be established. It has also become clear that the deeper the involvement the greater the number of errors which are uncovered. During the functional requirements review process, a number of high level problems were detected and corrected. However, it was not until the second team initiated their preliminary design that a significant number of other problems were identified. This suggests that the methodology should recognize the need for functional requirement modifications after some preliminary design and establish phased activities.

RSL/REVS is still an experimental tool which needs further improvements before it can be used successfully in a commercial environment. Specifically, computer resource requirements need to be reduced, and additional capabilities for hierarchical decomposition of a problem should be added.

In conclusion, the preliminary design phase has resulted in the detection and correction of significant errors in both the functional requirements and the preliminary designs. However, the cost effectiveness of the methodology employed cannot be established until the coding is done and testing is complete. At that point all detected errors will be analyzed and the effectiveness of the preliminary design effort assessed relative to more conventional industry practice.

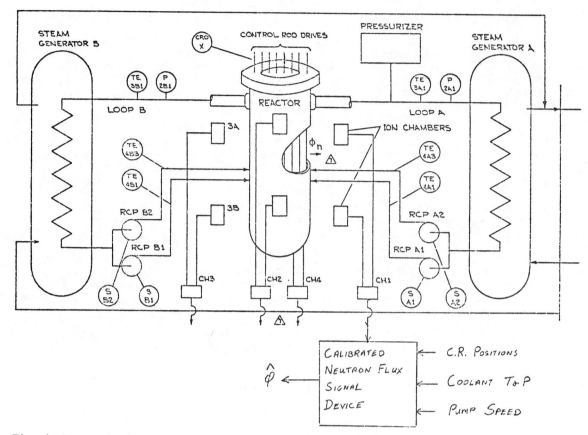

Fig. 1: Pressurized water reactor safety system instrumentation

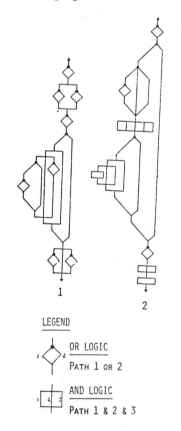

LEGEND

◇ OR LOGIC
Path 1 or 2

▭ AND LOGIC
Path 1 & 2 & 3

Fig. 3: RSL path enumeration
1 Team 1
2 Team 2

Experience with a specification language

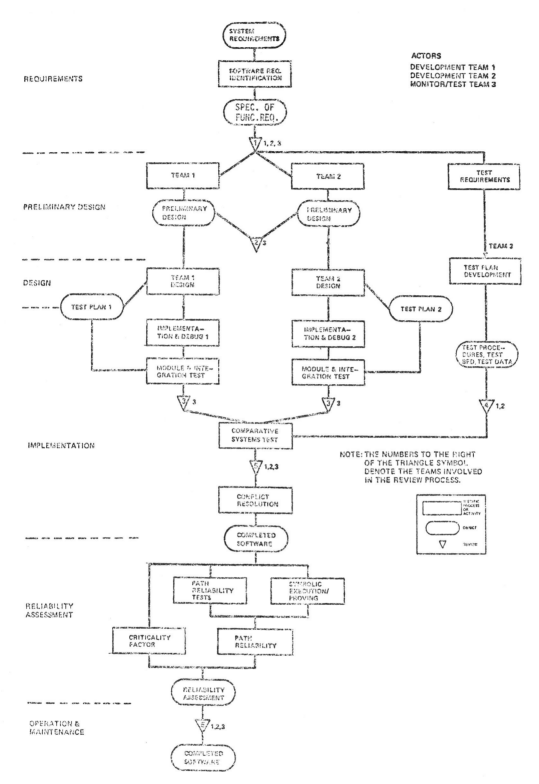

Fig. 2: Development methodology

OVERVIEW OF HARDWARE-RELATED SAFETY PROBLEMS OF COMPUTER CONTROL SYSTEMS

W. Schwier

Bundesbahn-Zentralamt, München, Federal Republic of Germany

Abstract. The main hardware failure types of computer systems are classified. The risks and the difficulties in the safety demonstration of different fail-safe computer systems are compared. Such computer systems are: Multi-computer with hard- or software comparator, with equal and diverse processing; single computer with checking program or with diverse multi-processing. The problems of computers with fail-safe circuits or with processing in coded form are discussed. A structure of the safety demonstration is recommended. In order to achieve a low probability of multiple failures a fast failure detection is necessary. In practise it is not possible to analyse the internal failures of a microprocessor in detail. An operational method of failure detection using a second processor and a function test program is outlined. Advantages and disadvantages of hardware diversity are discussed.

Keywords. Safety; computer hardware; microprocessor; fail-safe; failure modes; failure detection; hardware diversity; safety demonstration.

INTRODUCTION

Safe computer systems for the control or supervision of technical processes are needed if human life is endangered by the malfunction of these processes, if considerable material value could be destroyed or if the environment in which we live is being impaired.

If in this contribution, I speak of safety, I do not mean hazards directly originating with the computer, i.e., hazards in form of electric shocks caused by casings or operating devices being alive to earth, hazards through fire in the computer centre or hazards by an explosion called forth by an electric spark.

Our subject is the safety against hazards possibly developing in an automated process and produced by a malfunction of the computer. In these cases, by issueing incorrect control commands, the computer may cause a dangerous situation in the process; for example, if it switches a valve in a chemical plant incorrectly, this results in toxic gases becoming released, or if in railway operation, it switches a turnout incorrectly, this may result in the derailment of a train. In other cases, the malfunction of the computer system may lead to a situation where the system stops to perform the supervising- or protecting functions assigned to it. These incidents become hazardous if a danger situation in the supervised process is at hand and if the computer system - because of a defect - is unable to identify this situation.

FAILURE MODES

In the hardware of a computer, there are two types of failures to be expected which are to be distinguished by the time of their origin:

1 - failures which developed before operation or during repair work;

2 - failures developed during the operational phase in the computer hardware or in the peripheral circuits.

In the design of safe computer systems, it is most important to carefully distinguish between these two failure modes. Let us first look at failures developing during the operational phase.

As regards electronic components, the distinguishing features are stochastic and systematic failures. For stochastic failures it is possible to describe with sufficient accuracy the period between onset of load-condition and time of failure by using a probability distribution function. Stochastic failures occur statistically independent of each other.

Systematic failures originating during the operational phase are usually generated in other parts of the system where components have failed and are known as consequential failures. In other cases, we are confronted with regularly occurring failure mechanisms, i.e., failure mechanisms in form of a constant wear-out. As regards electromechanical components, this wear-out may in some cases be of importance. With computers it does not play a very important role.

A further distinguishing criterion are static and transient failures. The static failure, for example a short cut, remains unchanged after its occurrence. The transient failure is effective for a short period of time and vanishes again, but it may occur repeatedly. Causes for transient failures are electro-magnetic interferences or other influences, as for example vibrations, voltage variations, temperature variations.

Beside the single failure there are also multiple failures. It is inadmissible to apriori exclude multiple failures from the safety demonstration. In a poorly designed system, multiple failures may occur with a frequency which is comparable to single failures.

In the following, I shall attempt a qualitative comparison between the properties of different computer systems. For this, I shall class the failures in accordance with Table 1. As regards single failures, the table distinguishes only between static and transient failures. It is not necessary to distinguish single failures according to the cause (stochastic or systematic). The countermeasures are the same. In the case of multiple failures, the table distinguishes between the stochastic multiple failure and the systematic multiple failure including its categories "consequential failure" and "common-mode failure" which comprises failures developed before operation.

Unfortunately, there is not enough time available to deal with transient multiple failures because this topic would require a separate lecture. Transient failures have to be considered in combination with a static failure but they themselves also occur repeatedly, either in form of a burst in short successions or simultaneously in parallel working devices, or both.

COMPARISON OF SAFETY PROBLEMS OF COMPUTER SYSTEMS

Computer systems having safety functions may be classified as follows:

- multi-computer systems with comparison
  - equal computers with hardware-comparison
  - diversity computers with hardware-comparison
  - equal computers with software-comparison
  - diversity computers with software-comparison

- single computers
  - reliability
  - checking program
  - diversity multiple processing

- special designs
  - fail-safe circuits
  - processing in coded form

Table 1   Risk and difficulties in the safety demonstration

Failure		Multi computer system				Single computer		
		Hardware comparator		Software comparator		normal reliability	checking program	diverse multi processing
		equal	diverse	equal	diverse			
Single	static	◎	o	o	o	++	+	++
	transient	o	o	o	o	++	++	o
Multiple	stochastic	−	o	o	o	+	−	o
	consequential	−	−	+	+	++	++	++
	common mode *)	o	−	o	−	++	++	++

*) incl. manufactoring defects
◎ Reference point

o   equal   +   greater   −   smaller
                   ++ much greater   −− much smaller

Combinations of the listed systems are possible and customary.

Multi-Computer System with Comparision

This computer configuration is in widespread use if the safe control and supervision of a process is absolutely necessary. In this configuration, two or more computers work in parallel and perform the same task independent of each other. In this connection, "independent" means that failures occuring in one computer will not cause an unsafe change in the processing of one of the other computers. An evaluation logic is arranged between the elaborated signals and the process.

The first column of Table 1 shows a comparing evaluation of risks and difficulties which may occur with this system. Reference point for this comparison is the static single failure.

The transient single failure probably has the same influence on safety. The frequency of its occurrence is comparable to that of the static single failure. The computer system responds in the same way. In a multi-computer system, the stochastic multiple failure is dangerous only if its effects on several computers are of the same kind and if equally wrong signals - although for a short time only - are being produced simultaneously. The stochastic multiple failure is never to be neglected despite the fact that it is not as dangerous as the single failure. In multi-computer systems, it may be assumed that the relation between dangerous multiple failures and multiple failures in general is considerably lower than one. Therefore, in the table, the stochastic multiple failure was given the evaluation "minus". The risk it entails is smaller than the risk of a single failure. Treatment of the stochastic multiple failure in the safety demonstration is relatively difficult. I shall come back to this topic somewhat later.

A consequential failure is at hand if the failure of a component leads to failures of other components. The risk of the consequential failure is approximately equal to that of a single failure. On the other hand, in multi-channel systems and by use of decoupling measures, it can be controlled relatively easy and its treatment in the safety demonstration is also rather simple. Therefore, in the table, this failure was given the evaluation "minus".

Common-mode failures are at hand if several failures are produced by the same cause. They differ from consequential failures by the fact that the cause itself does not necessarily have to be a failure. Manufacturing defects, incorrect design, neglect of environmental conditions and wrong application lead to a simultaneous or almost simultaneous occurrence of comparable failures in several components. The evaluation of the common-mode failure is rather difficult. On the one hand and according to experiences, in proven designs it surely does not occur as often as the single failure. On the other hand, countermeasures are problematic if it concerns a system with similar multiple processing. If these two factors are carefully considered, the common-mode failure can be given the same importance as the static single failure.

Multi-Computer System with Diversity Processing

Above all, the problem of common-mode failures suggests the application of a diversity type of multi-computer system. Diversity may be related to software and also to hardware. At this point we shall discuss only the hardware problems. In this connection, we expect diversity to accomplish the following: Hardware failures which are not identified during preoperational tests should have different consequences in the two differently programmed processing units to enable the comparator to detect this situation. By the application of different hardware it may be possible to eliminate common-mode failures at all, either because only one of the computers will be at fault, or because the consequences will be different if both computers contain a failure.

Column 2 of the comparing Table 1 shows the risk and the difficulties which are to be expected in performing the safety demonstration for this system. Reference point is again the static single failure in the multi-computer system with equal computers. In the case of single failures there is no difference. In the same way, the behavior of the multi-computer system in the case of consequential failures does not change by the introduction of diversity. What amount of improvement can be expected for common-mode failures? Carefully estimating, we may expect the risk to become smaller. It is however, hard to tell if diversity will cause a considerable decrease of this risk.

In technique, it is very seldom that an advantage at one particular point may be achieved without a disadvantage at another point. Diversity substantially increases the cost for the system, i. e., the cost will approximately rise to double. Furthermore, diversity makes the fast failure detection which is performed by self-checking programs and by the comparator more difficult. Clock pulse synchronized operation with comparison at each step is not possible. The detection of failures can not be performed until - after a certain period of processing sequences - comparable states in both computers are reestablished. On the basis of these considerations, the table evaluates the stochastic multiple failure with a risk which is higher than in the case of systems without diversity.

Multi-Computer System with Software Comparision

It is possible to do without a separate hardware-type comparator. In such cases, the comparison functions are performed by the computers which contain comparison programs for this purpose. Each one of the comparison programs receives results to be compared from its own computer and from the other computer, as shown in Fig. 2. This process is performed by both computers. In case of discrepancy between the results, the comparing program stops its own computer.

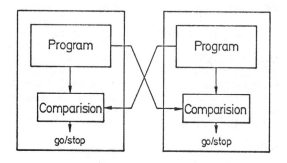

Fig. 2   Double Computer System with Software Comparision

We shall not discuss the question whether the multi-computer system with hardware-comparison or the one with software-comparison is economically more advantageous. We rather want to compare the risks and the difficulties to be expected in the safety demonstration. In the case of static and transient single failures, there will be no differences. The stochastic multiple failure will become less advantegeous because - similar to the case of diversity processing - the fast failure detection is more difficult. Consequential failures are more critical because, on the one hand, there is no decoupling of comparison- and user-programs in the computers which means that an error occurring in the computer may disturb the comparison as well as the processing. But above all, the necessary data exchange between the computers makes the proof of independence difficult. It has to be demonstrated that this data exchange is harmless even if a computer functions incorrectly. Finally, as regards common-mode failures, the system with software comparison will react in the same way as the system with hardware comparison.

Diversity Multi-Computer System with Software Comparision

With this system, the problems arising in the case of single failures are equal to those of the systems mentioned above. For common-mode failures, diversity is again an advantage. Failure detection and therewith stochastic multiple failures will not be more problematic than in the last mentioned systems.

Single Computers

So far we have discussed multi-computer systems serving for the safe control or supervision of technical processes. For economical reasons it was occasionally proposed to control or supervise a dangerous process at least temporarily by one computer. For this, there are different variants.

Reliable Single Computer

In the case of a single computer operated without special safety measures one has to rely solely on its reliability and on the fact that numerous failures will not become dangerous because the processing often stops after a certain period of time.

In principle, all failure modes shown in the table may become dangerous in a single computer. The situation of a static or transient single failure is particularly critical. The same applies to the consequential failures and to failures which developed before operation. Compared to single failures, stochastic multiple failures are somewhat less critical although no fast failure detection is provided in the normal computer.

Single Computer with Self-Checking Program

A self-checking program is arranged in the control system between the computation of output information and the output of signals to the process, as shown in Fig. 3. In the supervision system, this self-checking program runs cyclic. The self-checking program ascertains the correct functioning of the computer. If it detects a failure, it stops the computer. Output to the process is not possible unless the self-checking program has been run completely without detecting a failure.

How is this system to be evaluated? Firstly, to the present time, no self-checking program is able to detect every failure of a component contained in the computer. It is true that with conventional process computers it could be demonstrated that self-checking programs were able to detect nearly all failures in which the output signal of a circuit was set to high or low potential. It is however, difficult to master drift failures, time failures and failure combinations. For microprocessors this proof is also missing.

Let us attempt a comparing evaluation of risks and difficulties to be expected in the safety demonstration. The static single failure is in any case more dangerous than in multi-computer systems but not as dangerous as in a single computer without self-checking programs. The transient failure is very seldom detected by the self-checking program. If it affects only the user program, the self-checking program is unable to detect it. Its evaluation is as for single computers without self-checking programs.

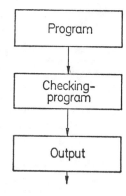

Fig. 3   Single Computer with Software Comparision

Consequential failures occur if a failure results in further failures. In this case, several components of the computer simultaneously function incorrectly. On the one hand, it is to be expected that the self-checking program detects a failure easier the more defects there are. But on the other hand, it is rather costly to examine failure combinations in the safety demonstration. For this system, the table evaluates the consequential failure with a "double plus". We recall the reference point for the comparing evaluation to be the single failure in the multi-computer system. The self-checking program is unable to detect failures which developed before operation; if it could detect them they would have been found already at the time of preoperational tests.

The situation with the stochastic multiple failure is similar to a multi-computer system. Here it is also important to detect a failure as fast as possible in order to make it ineffective.

With respect to this system, a further aspect has to be discussed. The period between the beginning of the self-checking program and the end of signal output is not to exceed a certain admissible time limit. Failures developing during this period can not be detected anymore. Although they will not be able to affect the processing, they may falsify the results already available or they may affect the output itself.

Single Computer with Diverse Multiple Processing

It may be considered to perform all processing work in a computer twice, one after the other. Figure 4 shows that all data and programs are doubled in the memory. The CPU first processes the data I with the program I and the result is being stored. Subsequently, program II ist running using the data II. Finally, a comparison program checks the two elaborated results with respect to their agreement in which case they are issued to the process. Because of the fact that a single failure would influence both processing operations in the same manner, a diverse processing is necessary. The design of both program systems and of both data sets has to be different.

It is very difficult to ascertain the degree to which static single failures can be made harmless by using diverse processing. This proof is very costly if the criteria applied for safety demonstrations is to be used. In this case it has to be shown for each individual single failure that it either affects only one program system or that it affects both program systems in such a different way that this fact can be detected by the comparison function.

With this system, i. e., with the single computer with diverse multiple processing, the transient single failure is not critical because it influences only one processing unit. Stochastic multiple failures may be evaluated in the same way as in the case of diverse multi-computer systems. Consequential failures and failures which the computer contains already at the time of first operation are as critical as single failures.

There are two further methods to be mentioned which are missing in the comparing table:

Computer with Fail-safe Circuits

In this case, the computer consists of fail-safe circuits. There were attempts to develop such computers but for economical reasons they were not successful. Because of today's availability of micro-processors which can be manufactured at low cost and in large numbers there seems to be no more chance for this solution.

Computer with Processing in Coded Form

The second method to be mentioned is the computer with processing in coded form. In communications technique, information to be transmitted are coded in redundant form enabling the receiver to detect or even correct transmission errors to some extent. If safety information is to be transmitted, the principle of coding is the same with the exception that there are higher demands on error detection. Furthermore, for coding and decoding it is necessary to use fail-safe circuits.

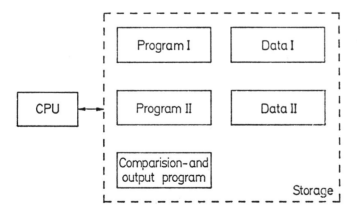

Fig. 4   Single Computer with Diverse Multiple Processing

It is common to protect information stored in computers, for example by a parity bit. If demands on reliability are higher, an error correction in semi-conductor memories is sometimes practised in order to prevent the failure of individual bits from resulting in failure of the memory.

It suggests itself to use this method for the failure detection in computers which are required to work in fail-safe mode. In this case, coded input signals are to be fed into the computer and processing has to be done in coded form. Results are being read-out and checked in a decoder.

Conventional computers or microprocessors are not designed for redundantly coded processing. The user may succeed in developing algorithms allowing the processing of coded data. It is however, not possible to provide instructions and addresses of the computer with an error detecting code. This would require a special computer designed for this task. It is therefore necessary to work with two separate programs.

This concludes the survey on different computer systems used for control and supervision of dangerous processes. The evaluation in the table has the purpose of enabling a qualitative comparison of computer systems. This comparison shows the critical nature of individual factors and the expenditures necessary for safety demonstration to cope with a given risk.

A criterion which is just as important are the costs. In this connection it is essential to detect the total cost. The simple method of counting computers is probably not sufficient. There are two cost factors to be pointed out: The software costs and the costs for the safety demonstration. Both easily exceed the hardware costs and in case of low piece numbers they determine the cost situation. With safety systems, the number of pieces is often rather small.

STRUCTURE OF THE SAFETY DEMONSTRATION

How can a safety demonstration be arranged? At least 4 steps are recommended:

(1) Single Failures

It must be demonstrated that no single failure may become dangerous. The basis for this demonstration is either the assumption that a component will fail nearly arbitrarily. Then, the course of the demonstration is as follows: If a component fails and if another component functions correctly, there is no danger. In the case of integrated circuits or of microprocessors, this is the way to proceed. Another possibility is the detailed analysation of a circuit. In this case, a list of the failure modes pertaining to components is necessary. For semi-conductor components, this list will be rather long and for a transistor, it may contain more than 50 failure modes. For integrated circuits, the elaboration of such a list would be possible only after analysation of the circuit which is practicable only in special cases.

Some properties of components have to be absolutely reliable because otherwise, it is not possible to design a safe control or supervision. The most important property is insulation. It is possible to design insulators, resistors, transformers and optocouplers so that they will suffer no short cut at the decisive points. In some fail-safe circuits, the hysteresis of magnetic material is being used. In these cases it is important that the coercive force - under certain conditions - is a lasting material property.

The basis with respect to single failures in safety demonstration is the knowledge of failure modes of a given component. It is the foundation for both the fault tree analysis (top-down approach) and the failure effect analysis (bottom-up approach). If this knowledge is not available, it must be assumed that the component will fail in nearly arbitrarily manner. What remains is the supervision of its function by another component.

(2) Independence

Just as important in safety demonstration is the independence. In the case of a fail-safe system, the components or units check each other in reciprocal manner. Premise for this is that no more than one unit becomes defective at one time. It would be dangerous if the failure of a unit would result in malfunction of the supervising unit, i. e., in consequential failures. Independence is for example necessary between two processing channels and the comparator. Safety for instance, is not available if one channel influences the other by falsifying a common input signal or if a failure in the comparator leads to modification of output signals of the processing channels. The proof of independence has to cover all coupling elements, for example, common ground wires, power supplies, spatial vicinity etc. It has to be demonstrated that there is no coupling or that the coupling remains undangerous. Decoupling is best achieved by the aforementioned insulation components, namely resistors, transformers, optocouplers, and by sufficient spaces between conduction paths.

Every form of information transmission between the processing channels is a potential danger because it may introduce errors into the other branch. Pure start- or semaphore-variables are also risky because the output of a correct signal at the wrong time may be dangerous. If one computer - because of an error - starts a program at the wrong time, there is the danger that by presentation of a start signal in the other computer the matching program will also be started. The comparator would not be able to detect this error. The same applies to the absence of start signals which is caused by a failure.

(3) Multiple Failures and Failure Detection

The next point to be discussed with respect to safety demonstrations are stochastic multiple failures. These are failures occurring within the same period but having independent origins. Basically, there are two possibilities to cope with stochastic multiple failures. One possibility is to check if the multiple failures are dangerous; the other possibility is to make the probability of their occurrence so low that they need no further attention. In practice it is common to choose the second possibility; only if a stochastic multiple failure can not be made improbable, the failure combinations are being checked but expenditures for this are very high.

The probability of a stochastic multiple failure becomes small if the first failure is detected in sufficient time and if the affected component is switched off to the fail-safe side. A failure may be self-signalling, i. e., on its occurrence it triggers a response resulting in the switching-off. But the failure may also be detected by the data flow. In this case, failure detection time is dependent on the data flow which is determined by the incidents in the process. The failure detection time varies within wide limits.

If the data dependent failure detection times are too long, it is necessary to apply special checking programs which are to detect the failures and which stop the computer directly or indirectly by changing the output signals with the aid of the comparator. The decisive problem is: How can one determine if the checking program detects a developed failure? A microprocessor - the same applies to a computer with LSI circuits - can not be disassembled. The connection diagram of its components is not sufficient for analysing the effects of failures. It is rather necessary to also consider the embedding of components in the substrate. The arrangement of components has to be known in order to recognize short cuts between adjacent components. How are conducting paths arranged? Which components may be affected by short cuts? For the evaluation of the effects of interruptions it is necessary to know the enbranchments of conduction paths. The proof of efficiency of self-checking programs which bases on failures will be very costly and is probably not practicable. But in any case, it is valid only for one design of a microprocessor. Manufacturers occasionally improve the lay-out of integrated circuits or the manufacturing process without changing designations or data sheets. In such a case, the demonstration would have to be repeated.

Another approach which may be termed as operational method will perhaps bring a solution. If all functions of a control or supervision are performed at least once within the time granted for the failure detection and if two parallel working microprocessors are checking each other in reciprocal manner it is possible to tell after completion of a cycle if each one of the microprocessors fulfills its function. It is in this case absolutely possible for one register or one instruction to be defect. If this fact has no effect on the functions demanded during this application, the situation will not become dangerous even if the other microprocessor should also fail.

Unfortunately, in practice there are few processes in which all functions are performed regularly within a short period of time. It is however, possible to add further program steps in form of an on-line function test program which periodically requires the functions of microprocessors. By comparing both processors it may be checked if the functions are still in agreement. In the affirmative case, they are considered to be free of errors. Basis for the design of the function test program is a functional model of the microprocessor containing all parts which could possibly be addressed by the instruction set. The model components have to work according to the examined description of instructions, i. e., model and real microprocessor completely agree in those functions which are identifiable from outside.

If - after completion of the function test program - the comparator detects a malfunction in the model, it is to be hoped that it will detect such malfunctions in the real processor also. In the industrial field and in traffic, this detection of failures should be sufficient for making stochastic multiple failures fairly improbable.

This approach assumes the microprocessor to be a system with known structure. The structural elements are described with respect to operation. The operations to be performed by the elements are reactions to instructions or are storing processes in registers. For safety demonstration, the structure has to be divided in such a manner, that the reaction on each instruction can be located. The comparison of two microprocessors determines the identity of two performed operations. In the affirmative case, there is a large probability that the functions demanded by the process are also being performed orderly.

With this approach, it is important to compare two equal microprocessors during each individual operation. If necessary, the result of one operation has to be read out by means of storage instructions so it may reach the comparator.

Failure detection through the comparator is most efficient if all input- and output-signals of processors are directly compared during every clock pulse, i. e., if not only output signals of a computer but also data- and address busses as well as control signals are being compared. If only output signals of complete computers are being compared, this failure detection loses much of its efficiency. In this case, the function test program has to be much longer and more complex because there is the additional task of transporting the internal results of operations to the output pins to compare them. Now, failure effects may cover each other in subsequent operations. It is necessary to additionally demonstrate that this situation will not arise.

Accordingly, there is the following dilemma at hand: In the first case, more hardware by the addition of a comparator which has to grasp all input- and output signals with every clock pulse. In the other case, more expenditure in the memory and a higher demand in time for the function test program and more expenditure in the safety demonstration.

(4) Safe Disconnection

The fourth point in a safety demonstration is the safe disconnection. It should be demonstrated that a detected failure results in disconnection which can not be cancelled by the occurrence of further failures.

HARDWARE DIVERSITY

In order to render common-mode failures or failures which developed before operation undangerous, it was proposed to include diversity in parallel channels the results of which are to be compared. The main reasons for this are problems occurring during a program system verification through which the system is proven to be free of errors. This paper treats only hardware failures, i. e., advantages and disadvantages of diversity in hardware.

If both computers contain an error it is unquestioned that a system with two diverse processors is safer than a system with two equal processors. Must it be assumed that two new and equal microprocessors contain the same error not detected during the check-outs performed before operation? A check-out completely independent from application is not possible. More successful seems an application dependent check-out during which user programs are performed as completely as possible before operation. This check-out will not detect everything and it should therefore be performed in combination with a second measure: Both microprocessors should not come from the same manufacturing process. If they are the product of one manufacturer, best results are achieved if they come from different production lines and if there is a somewhat longer interval between their manufacture. There is the further possibility (or this possibility may be managed) to observe the efficiency of microprocessors of one series during applications requiring no safety. For safety applications, they are to be used only after a certain period of observation time.

A substantial technical disadvantage of diverse microprocessors are the difficulties experienced during failure detection. It is no longer possible to detect failures by constantly comparing two equal processors. In the case of different processors, each one of them has to detect its own failures by a self-checking program and - in a given case - has to stop its own operation. This is a task which is rather difficult or even impossible to accomplish because the criterion provided by the other processor is missing. The application of different or differently programmed microprocessors substantially increases the costs for a given system.

These additional costs are a factor which certainly influences safety. Safety costs money and it is necessary to spend the financial means available in such a manner that they yield a maximum in safety.

DISCUSSION

Ryland: We have some experience with multi computer systems with software comparison. In the systems we have built, we have connected the outputs of each processor directly as inputs to the other processor. With this in mind, I would like to make two points about consequential failures. First, a failure which causes both processing and comparison errors in one processor is acceptable because any resultant error in the output will be observed by the other processor and the system will shut down. Second, the data exchange between the programs is reduced to synchronization and input confirmation signals. The program can be written in such a way that errors in these exchange will either have no effect or will cause a processor shut down.
I would like to ask Mr. Schwier whether he would agree that the system I have mentioned is different from the one he described and thus avoids some of the problems of consequential failures.

Schwier: I don't see a fundamental difference to the principles of the computer system with software comparison, I spoke of. I think the main difficulty with this system is that you have to feed signals from one computer into the other. If these signals have a fault concerning the logical or the physical form, then it is possible, that the other computer will also work in a wrong way. I said in my lecture that it is more difficult to demontrate the safety in case of consequential failures compared with the system having a hardware comparator.
The next problem is to shut down the system. If the comparison program detects a failure then you must be able to shut down the computer. But you may have a failure, so that the comparison program is not able to do this.

Ryland: If you get the failure in one processor, so he can get the incorrect output, its up to the other processor to do the shuting down the process. So about both the processing and the comparison task the program running on one processor are affected by a failure, it does not influence the safety of the system. The comparison program will run on the other processor to shut the system down.

Schwier: I am not sure that we will be able to overcome this difficulty. I try to point out the difficulties in the systems I know. It is quite possible, you found a way to solve this problem in a safe way. The comparison program in the fault free computer stops its own computer. But you will have the difficulties in the other computer, which may not be able to stop due to the failure. How will you block the wrong output signals of this computer without a hardware logic? And it is also possible that due to a hardware failure the comparison program alters one of the results. In this case both computers have different outputs. You can provide decoupling measures by resistors or optocouplers or by other means so that there is no failure propagation to the results. But you can do this only with a special fail-safe hardware not with programs.

Grams: Mr. Schwier, one question to your table, where various types of systems have been compared. I am interested in the background leading to these conclusions. Have there been studies and evaluations based on realized systems or have there only been preliminary assessments of proposed systems?

Schwier: Most of the systems have their background in implemented systems I know. The equal multiprocessor systems with hard- or software-comparator, also the systems with diverse processing. We have no multiprocessor system with diverse hardware. Then we have the single computer (for short times) with diverse multi-processing. A fail-safe computer was designed some years ago in a foreign country, but as far as I am informed, it was only set into operation parallel to a relay system. In our country the computers with coded processing are only proposals - there is a new proposal I heard in the last time.

I tried to make up clear separation of the principles therefore I simplified in some way. The real systems are mixtures.

Grams: Which existing systems have been examined?

Schwier: In detail systems from German manufacturers, from Siemens, SEL, AEG.

THE COMBINED ROLE OF REDUNDANCY AND TEST PROGRAMS IN IMPROVING FAULT TOLERANCE AND FAILURE DETECTION

H. Schüller and C. Santucci*

Gesellschaft für Reaktorsicherheit mbH, Forschungsgelände, 8046 Garching, Federal Republic of Germany
*Società di Elettronica per l'Automatione S.p.A., Lungo Stura Lazio 45, 10156 Torino, Italy

Abstract. The reliability and safety of computer control systems is becoming steadily more important. Beside the logical correctness of the programs their fault-free performance must be ensured and supervised. Therefore hardware measures, which are suitable to facilitate this, are of fundamental importance.

Due to the hardware aspects two aims should be reached:

1. Prevention of failure effects from the process, i.e. fault tolerance.
2. Rapid and complete failure detection during operation.

Fault tolerance may be obtained by redundancy, i.e. by a suitable system structure. However, for rapid and complete failure detection in a computer test programs must be developed. Normally both methods will be applied in control systems used with safety responsibility.
So on the one hand different possibilities of system structures are studied with respect to improvement of safety and on the other hand possibilities and bounds of test programs are analysed in relation to rapid and complete failure detection. Some hardware features to increase effectiveness of redundancy and completeness of test programs are also described.
Mathematical relations are derived for quantitative determination of figures of merit. This relations show the influence of failure rates, rapidity and completeness of failure detection, repair rates and degree of redundancy in evaluating reliability and safety.
It is shown that

1. Redundancy can make important improvements on the failure detection in computers, and
2. rapid and complete failure detection makes large increases of the probability for fault tolerance by redundancy.

Thus, by combination of both methods, the system reliability as well as the system safety will be improved.

Keywords. Computer Safety, Fault Tolerance, Test-Programs, Failure Detection, MvN-Systems.

INTRODUCTION

Lately tasks involved with safety responsibility are increasingly being taken over by process computers and microprocessors /1,2/. Thus the question "what happens if the computer system fails" becomes more and more important. Such failures can be caused by software errors as well as by hardware faults. A distinction between both is often difficult. E.g. an undesired change of the content of a memory cell can be due to a software error as well as due to a hardware fault. This paper handles only with hardware faults. Its definition shall come from the failure cause, i.e. that any effect which comes from a hardware fault is appointed to the latter. In the above mentioned example an undesired change of the content of a memory cell, which could be interpreted as a software fault will be associated with a hardware fault if the latter was the original cause.

A priori it is clear that dangerous effects coming from a hardware fault in a safety system must be avoided. In order to reach this aim two possible methods can be used:

a) quick failure detection and automatic reaction to the safe side in case a failure has been found.
b) prevention of incorrect output signals from the process.

In case a) failure detection can be reached by retest, test programs, circuital test tools, or similar methods. Sometimes a very elegant possibility exists to make use of special features of materials of constructions in order to get automatic failure detection and reaction to the safe side in case a hardware failure occurred: the fail-safe technique. But up to now process computers and microprocessors can not be built in this fail-safe technique. For this reason special failure detecting methods are needed such as test programs, which are very suitable for computers and microprocessors. However, the following control mechanism - the connecting unit between the computer and the process - must be built in fail-safe technique. Furthermore it must be mentioned that incorrect control signals must be tolerable by the process for a certain period (until reaction to the safe side after failure detection).

In case b) one tries to prevent incorrect output signals from influencing the process. This can be done e.g. by multiple i.e. redundant calculation of all necessary i/o associations and by forwarding output signals to the process only in those cases where they were found to be unequivocal or by a majority, respectively. The majority decision logic has to be built in a fail-safe technique as is necessary for the connecting unit to the process in case a).

STATE DIAGRAMS FOR SINGLE SYSTEMS

The simplest case for a state diagram of a not-repairable system is shown in fig.1.

Fig.2 and 3 show examples for the probabilities P_i of being in the different system states calculated for special values of λ and ε. One can see the strong influence of ε to P_2.

If it is possible for a system to switch from the failure-free state into different failure states by (different) failure rates λ_i, one has first to clarify whether all these transients are possible from one original state as shown in fig.4 or whether the original state has to be splitted accordingly, because each of the different failure modes lead to a distinct failure state as shown in fig.5.

For system failures caused by defects of electronic components the second case (fig.5) is decisive. If the different failure modes are all completely independent from each other (and if common mode failures must not be considered) the probability P_s for the system being in the failure-free state can be calculated as follows:

$$P_s(t) = \prod_{i=1}^{n} P_i(t)$$

(P_i = probability that the subsystem i is failure-free)

The state diagram of fig.5 can be enlarged according to fig.6 if failure detection is introduced. Considering also repair fig.7 results.

For the stationary case and infinitesimally quick repair the expression for the availability e.g. results in

$$A = P_s(t\to\infty) = \prod_{i=1}^{n} \frac{\varepsilon_i}{\lambda_i+\varepsilon_i}$$

Thus all probabilities for the popularities of the states of single systems can be calculated even for several failure modes and by considering failure detection and repair. Now we try to calculate those probabilities for M out of N (MvN) systems. The nomenclature of an MvN system shall be derived from a protection system /5/.

STATE DIAGRAMS FOR M OUT OF N SYSTEMS

When we regard redundant systems we must first differ whether system repair is possible or not. If we need a system only at special instants as is the case e.g. for a reactor protection system, then also the (dangerous) system failure, is repairable, because it is only important that the system is intact at the time of demand. If on the other hand a system is needed continuously as it is the case e.g. for a flight control system, then system repair is not possible, because the (dangerous) system failure leads directly to the undesired event. But also in the latter case the redundant components of the system are repairable.

Fig.8 shows the state diagram of a repairable MvN system, while in fig.9 the state diagram of such a system is shown for the case that the (dangerous) system failure is not repairable. This leads to so-called absorbing states.

By solving the corresponding differential equations the population probabilities for the different system states can be calculated time dependent and quantitatively. Through this way one can get figures of merit (probability values) for the system reliability and the system safety by summing up the population probabilities for all reliable system states, or safe system states respectively. While those figures of merit can be calculated only numerically for systems, which are not repairable they can be gained also analytically for repairable systems /6/.

For MvN systems e.g. we get:

$$A_{MvN}(t) = \sum_{k=M}^{N} \binom{N}{k} A(t)^k \bar{A}(t)^{N-k}$$

with $\bar{A} = 1-A$ = unavailability of a single component.

$$MTBF_{MvN} = MTBF \frac{\sum_{k=0}^{N-M} \binom{N}{k}(\frac{\bar{A}}{A})^k}{M\binom{N}{M}(\frac{\bar{A}}{A})^{(N-M)}}$$

The same formulas are applicable for the safety function for demand operation /3/. We have only to change A into S_d and MTBF into MTDF (mean time between dangerous failures). For a 2v3 system e.g. we get:

$$\text{MTBF}_{2v3} = \frac{3(\frac{\lambda}{\varepsilon} + \frac{\lambda}{\mu})+1}{6\lambda(\frac{\lambda}{\varepsilon} + \frac{\lambda}{\mu})} \quad ; \quad \text{MTBDF} = \frac{1}{6}(\frac{1}{\lambda} + \frac{1}{\mu})(\frac{\varepsilon}{\mu} + \frac{\varepsilon}{\lambda} + 3)$$

THE INFLUENCE OF REDUNDANCY AND OF FAILURE DETECTION ON SAFETY FIGURES

The mathematical influence of failure detection and redundancy on safety figures has been derived in the previous chapters. The probability for a system failure can be decreased also by pure increase of redundancy without failure detection and repair (fig.10).

Without special failure detection even a 2v3 system is much too unreliable for safety applications as shown in fig.11 /7/. The combination of redundancy with quick failure detection and repair, however, diminishes considerably the system failure probability (fig.12) /7/.

The question how to built up suitable redundancy is important but easilier solved /8,9/ than the question of how to reach quick and sufficient failure detection.

One method of rapid failure detection in computers is the use of self-checking programs. Such test programs have been written e.g. in an on-line version for an AEG 60-10 process computer /4/. Their essential parts are:

A) Special function test programs:
   a) instruction test program
   b) core store function test program
   c) I/O function test program

B) global supervision programs
   a) core store constant data test routine
   b) program flow monitoring routine.

These programs are running partly in a fixed cycle in the foreground, (A.a.,A.c.,B.b), partly in the background within the free cpu-time (A.b.,A.c.,B.a.). The whole test program system is used on-line within the reactor protection computers which are installed open loop at the nuclear power plants in Brunsbüttel and Philippsburg. They guarantee that component failures are detected and brought to the safe side within either 200 msec (foreground test) or within 60 sec (background test).

Naturally the failure detection of these test programs has to be verified. To do this the output pins of all gates were forced statically to 0 volt and +5 volts respectively (stuck-failure). This was possible because this process computer, is not yet built with highly integrated circuits. This practical test was performed during normal program run, and the intervall was timed from failure introduction to failure detection by the fall of pulse supervision relays which are connected to the computer outputs.

SPECIAL RESULTS

It was found that

a) the last version of the self-checking programs detected all simulated failures as far as they could influence the program run at all,

b) about 97,5% of all simulated failures were turned to the safe side within 200 msec and the rest within 60 sec.

Of course we must always consider that a certain part of all possible failures cannot be detected by the test programs. Such failures will be determined only after corresponding demands from the process (special input data combinations to the computer). But in redundant systems those failures are still without influence on the process.

Thus it is obvious that the combination of data independent failure detection by self-checking programs, and data dependent failure detection by redundancy is most favourable for increasing reliability and safety figures. Both methods intensify their improving effect on these figures and complement each other. This is made evident in the following example (fig.13, /3/).

According to the mentioned example of a special computerized 2v3 reactor protection system the results shown in fig.14 could be reached.

LITERATURE

/1/ Büttner, W.E. (Juli 1976). Sicherheitstechnischer Vergleich eines festverdrahteten dynamischen Reaktorschutzsystems mit einem Rechnerschutzsystem (Teil 1). Laboratorium für Reaktorregelung und Anlagensicherung, Garching, Lehrstuhl für Reaktordynamik und Reaktorsicherheit, MRR-Bericht Nr. 161.

/2/ Strelow, H., and H. Uebel (1978). Das sichere Mikrocomputersystem SIMIS. Signal und Draht 70, 4, April, S. 82-86.

/3/ Plögert, K., and H. Schüller (1977). Process Control with High Reliability; Data Dependent Failure Detection Versus Test Programs. 5th IFAC/IFIP Int.Conference on Digital Computer Applications to Process Control. The Hague, The Netherlands, June 14-17, Proceedings pp 695-703.

/4/ Schüller, H. (1978). Methoden zum Erreichen und zum Nachweis der nötigen Hardwarezuverlässigkeit beim Einsatz von Prozeßrechnern. Dissertation, Technische Universität München, Institut für Automationstechnik.

/5/ Schüller, H. (1977). Different Interpretations of the M out of N Systems with Safety Responsibility. Purdue Europe, TC7 Safety and Security, WP No. 120

/6/ Applebaum, S.P. (1965). Steady-State Reliability of Systems of Mutually Independent Subsystems. IEEE Transactions on Reliability, March, pp 23-29.

/7/ Schüller, H. (1974). Application and Functional Test of Self-Checking Programs; their Influence on the Failure Probability of Computerized Safety Systems. Enlarged Halden Programme Group Meeting on Computer Control and Fuel Research to Safe and Economic Plant Operation Sandefjord/Norway June 3rd-8th.

/8/ AEG-Leistungsreaktoren (1973). Abt. E316 Reaktorüberwachung. Rahmenspezifikation für Reaktorschutzsystem mit Prozeßrechnern. KKB FC/BA 005 vom 2.1.73

/9/ Wobig, K.-H., A. Hörder, and H. Strelow (1974). Prozeßrechnersysteme mit Fail-Safe-Verhalten. Signal und Draht, 66, 11, ppa. 211-218.

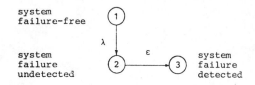

λ = failure rate
ε = failure detection rate

① safe and available state

② not safe and not available state

③ safe but not available state

Fig.1: State diagram of a not repairable system.

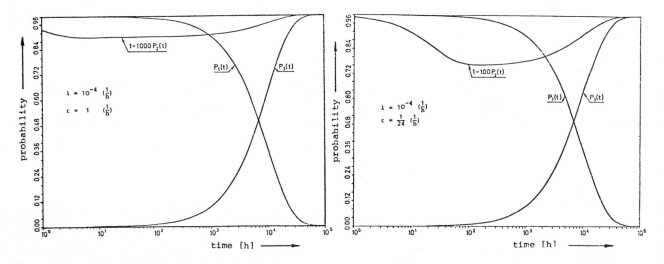

Fig.2: Probability distributions P_1, P_2, P_3 corresponding to fig.1.

Fig.3: Probability distributions P_1, P_2, P_3 corresponding to fig.1.

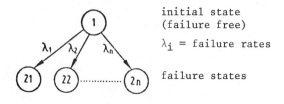

initial state (failure free)

λ_i = failure rates

failure states

Fig.4: State transitions from one initial state into different failure states.

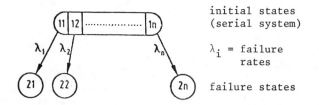

initial states (serial system)

λ_i = failure rates

failure states

Fig.5: State transitions from several initial into corresponding failure states.

Redundancy and test programs 183

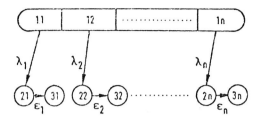

Fig.6: State diagram for different failure modes including failure detection.

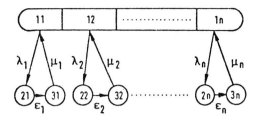

Fig.7: State diagram for different failure modes including failure detection and repair.

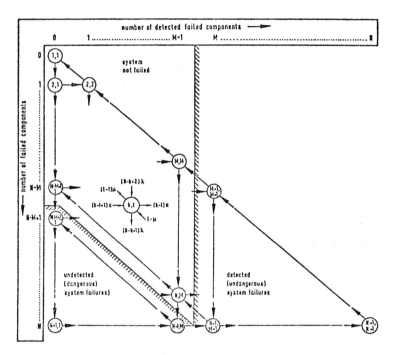

Fig.8: State diagram of a MvN-system; all detected failures, even system failures, are repairable.

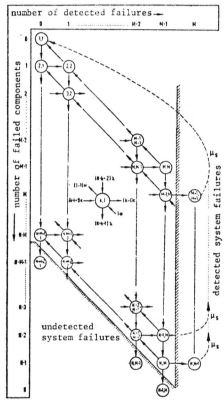

Fig.9: State diagram of a MvN-system. Undetected system failures are absorbing states. Detected system failures can be absorbing states, otherwise μ_s=system repair rate.

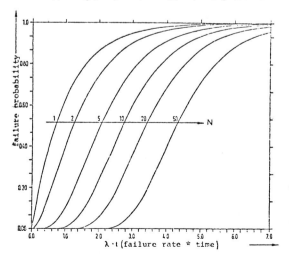

Fig.10: Failure probability of an 1vN-system.

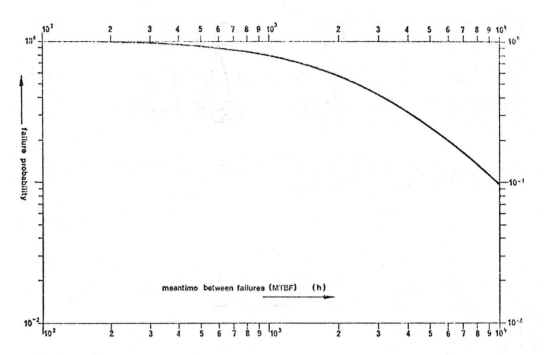

Fig.11: Failure probability in case of a shutdown demand for a computerized 2-out-of-3 protection system dependent upon the MTBF of a single computer, if not special precautions for rapid failure detection are taken.

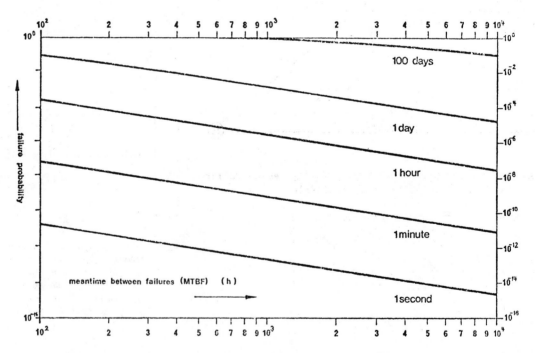

Fig.12: Failure probability in case of a shutdown demand for a computerized 2-out-of-3 protection system dependent upon the MTBF of a single computer, if a limited detection time for all component faults exists which is the parameter in this fig.

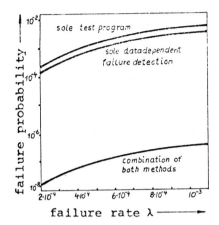

Assumptions for the example in fig.13:

Mean test program cycle time: 60 sec.
Completeness of failure detection: c=0,99
(the remaining with data dependent failure detection or within 1 year).

Data dependent failure detection:
   80% within 5 min.
   80% of the remaining within 1 day
   80% of the remaining within 1 week
   the remaining within 1 year.

Mean time to repair: 1 day.

Fig.13: Failure probability for dangerous failures of a 2v3-system for two failure detection methods.

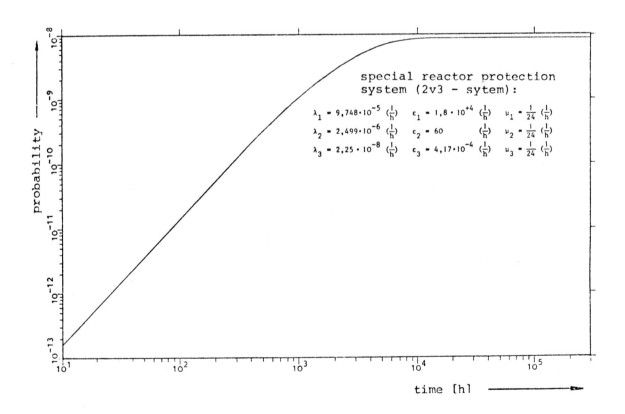

Fig.14: Time dependency of the maximum probability for undetected (dangerous) failing at a demand of the special reactor protection system.

A FAIL-SAFE COMPARATOR FOR ANALOGOUS SIGNALS WITHIN COMPUTER CONTROL SYSTEMS

G. H. Schildt

*Engineer for Safety Technique, Member of VDI-GMR and VDE-NTG,
D 3300 Braunschweig, Federal Republic of Germany*

Abstract. The discussion of different concepts of comparators for analogous signals within computer control systems with responsibility for safety is followed by the description of an intrinsically safe comparator.

Keywords. Fail-safe comparator; tolerance zone; different principles of comparators; fail-safe relais; multi-channelled system; design principle of diversity

INTRODUCTION

In the field of safety technique redundant control systems for processes with responsibility for safety are often used, because single-channelled and fail-safe control often is not possible at all or only at high technical costs which cannot be justified. By using multi-channelled non-safe control systems the responsibility for safety is passed over to a comparator to be installed at the outputs of each control channel. The development of a single-channelled, safe comparator for analogous voltages in the operating range of approx. 10 mV to approx. 15 V is especially difficult, because in addition to the observance of a given tolerance zone - depending on the voltage to be compared -, responsibility for safety of the comparator has also got to be taken into account. Thus comparators for analogous signals are to compare safely two input voltages U_1 and U_2 according to fig. 1 and, from the result of this comparison derive a decision:

log.'∅' non-equivalent
log.'1' for equivalent

according to a given tolerance zone

In this connection a tolerance zone means the curve of the voltage difference $\Delta U = U_1 - U_2$ versus the voltage range U' of the voltages to be compared, as shown by fig. 2a. The ideal comparator for analogous voltages (in the following called 'analog comparator') shows a tolerance zone dependance according to fig.2a.

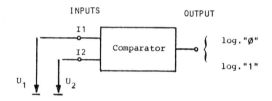

Fig. 1. Comparator for analogous voltages

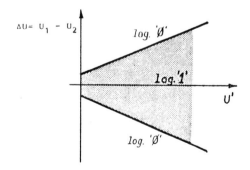

$\Delta U = f(U')$
with $\Delta U \neq 0$ for $U' = 0$

Fig. 2a. Tolerance zone versus U'

The analog comparator shows an ideal tolerance zone behavior if, for

$$\left|\frac{\Delta U}{U}\right| = \text{const},$$

the tolerance zone opens up, symmetrically and linearly, to both sides of the ordinate, with the condition that $\Delta U (U' = 0) \neq 0$.

REQUIREMENTS FOR ANALOG COMPARATORS RELEVANT TO SAFETY

It is an essential characteristic of a safe analog comparator that safe '∅'-information occurs at the output, in case of unequality of the input variables to be compared, as well as in case of failure of single or multiple structural elements. This corresponds to the following requirements from the technical point of view for safety:

- safe comparison of $U_1 = \emptyset$ with $U_2 = \emptyset$,

- maintenance of the tolerance zone dependance in case of failure for $|U_{1,2}| > \emptyset$ without rotational or translatory shift of the tolerance zone (potential failures, according to fig. (2b), have got to be safely controlled),

- safe '∅' at the output of analog comparator with open inputs (e.g. resulting from broken wires or detached connecting plugs),

- for a multi-channelled comparator system the mastery of drift-effects and commonmode disturbances,

- defined temporal cut-off behaviour of the comparator configuration,

- to guarantee the tolerance zone within the large as well as the small signal range,

and

- electromagnetic compatibility (emc).

Shifts of the tolerance zone are not dangerous, if the 'old' and the 'new' tolerance zone do not overlap within the operating range, defined by $U'_{min} \leq U' \leq U'_{max}$ (see fig. 2c), because such a shift immediately will cause a cut-off reaction.

But if the old and the new position of the tolerance zone overlap within the fixed operating range of the voltages to be compared - caused by drift-like or jumping changes of the tolerance zone - then, however, a danger may issue from the comparator, as shown by fig. 2d.

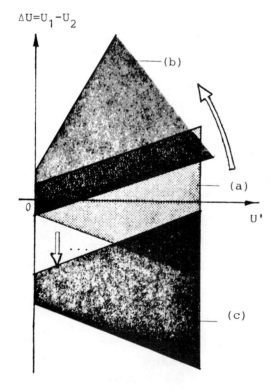

Fig. 2b. Tolerance zone
- in normal position (a)
- after rotatory shift (b)
- after translatory shift (c)

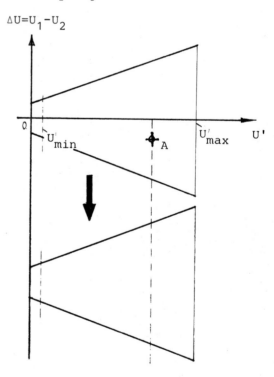

legend: A = operating point with U_A within the admissible tolerance zone

Fig. 2c. <u>Undangerous</u> tolerance zone shift

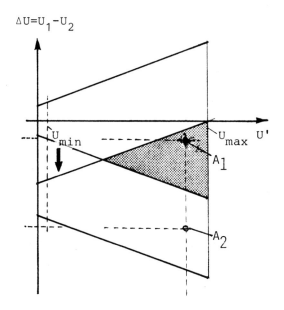

legend: A=operating point with U_A

Fig. 2d. Dangerous tolerance zone shift

For comparators relevant to safety various hardware concepts are conceivable. They shall be separately discussed in the following.

VARIOUS CONCEPTS OF ANALOG COMPARATORS

Fig. 3 shows various principles how to design analog comparators. A fundamental difference is made between concepts w i t h o u t and w i t h secondary energy. Furthermore a difference is made between comparators operating with static or dynamized signals.

The different concepts will be discussed in the following.

ANALOG COMPARATORS WITHOUT SECONDARY ENERGY

The fundamental requirement to be fulfilled by a safe analog comparator with a safe logical 'Ø' at the output can be met by designing a device without secondary energy (i.e. without additional supply voltage). Again, concepts according to fig. 3 can be differentiated: One group is run by electronic static signals and the other by electronic dynamized signals. Such concepts, however, have a special limitation:

The energy necessary for the comparison of the input variables U_1 and U_2 has got to be derived from the input variables themselves. If according to fig. 4, one assumes a certain

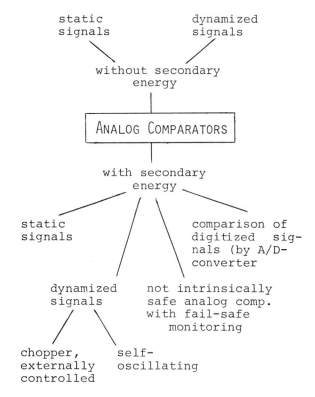

Fig. 3. Concepts of analog comparators

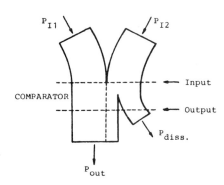

$P_{I1} + P_{I2} = P_{Out} + P_{diss.}$;
Essential condition for 'cascading', if necessary:

$P_{Out} \stackrel{\geq}{=} (P_{I1} + P_{I2})/2$

legend: $P_{I1,2}$ = power of the input signal
$P_{diss.}$ = dissipation
P_{Out} = output power

Fig. 4. Balance of power for an analog comparator without secondary power

power dissipation $P_{diss.}$ of the comparator device, an output power

$P_{Out} \geq (P_{I1}+P_{I2})/2$ is needed for cascading. An important advantage of a comparator concept without secondary energy is given by the relief to get a safety proof: A safe 'Ø' at the output can be guaranteed (e.g. at broken input wires), because no uncontrolled oscillations may occur within the comparator device.

Concepts for comparators using static electronic signals have got to be rejected, because it can never be guaranteed that - due to a failure of an electronic constructional element - there is permanently the information log. '1' for 'equivalent' at the output of the comparator device. Therefore concepts with dynamized, electronic signals are recommended for analog comparators relevant to safety.

ANALOG COMPARATORS WITH SECONDARY ENERGY

The next group of analog comparators is the one using secondary energy (fig. 3). The fact that these comparators are operating with secondary energy leads to an additional restriction, because in case of failure, one has to proof that a log. '1' only caused by uncontrolled oscillations within the device is impossible at the output. According to fig. 3 an additional difference can be made between comparators containing

- a chopper, driven by an external oscillator

and

- a self-oscillating system.

COMPARATOR WITH A CHOPPER CONTROLLED BY AN EXTERNAL OSCILLATOR

Fig. 5 shows the fundamental block diagramm of an analog comparator with choppers controlled by an external oscillator, based on an application for a patent according to /2/. In order to obtain sufficient properties of the tolerance zone in the small signal range as well as in the large signal range, this proposed concept shows a double-channelled structure. The term 'double-channelled' exclusively refers to the function, and not to the safety function.

The input voltages to be compared U_1 and U_2 are dc-voltages; they are dynamized by externally controlled choppers and the sum of $(U_1 - U_2)$ plus the chopper signal is compared with 'Ø'. Thus in case of equality to the voltages ot be compared energy will be transformed and rectified

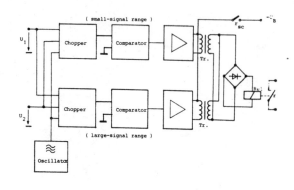

legend: r_{sc} = self-killing contact
  r = comparator-contact, potential-free
  Rel = signal relay
  $+U_B$ = supply voltage, secondary energy
  $U_{1,2}$ = input voltages to be compared

Fig. 5. Comparator with choppers controlled by an external oscillator

so that the output relay attracts. If, during the operation, unequality of the voltages to be compared occurs, then the relay drops, and according to the closed-circuit-current-principle the contact r opens. At the same time secondary energy supply is separated from the comparator device by the 'self-killing-contact' r_{sc} according to the 'self-killing-concept'. This guarantees that, in case of a recognized unequality of the voltages to be compared, a new attraction of the relay Rel is impossible. It has to be pointed out for this concept that it has to be proved that in this system no uncontrolled oscillations may occur, although this system is not a feedback system.

SELF-OSCILLATING COMPARATOR

The principle of a self-oscillating comparator is based on the patent application according to /3/. It is shown in fig. 6.

The input voltages U_1 and U_2 are compared with each other in such a way that in case of equality in the feedback-system - consisting of an amplifier with amplification v_u and feedback with the coupling coefficient k - oscillations are excited with the frequency f. Their energy, subsequently rectified, causes a relay Rel to attract. As, however, this feedback system is, in principle, a circuit

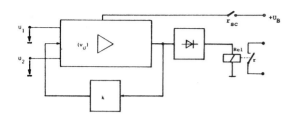

$k \cdot v_U = -1$ for $U_1 = U_2$

legend: r_{sc} = self-killing contact
$+U_B$ = supply voltage
k = feedback-factor

Fig. 6. Self-oscillating comparator

design which provides for oscillation capacity from the beginning, one runs into difficulties to get a safety proof, because it is difficult to prove, that no other oscillations of deviated frequencies may occur. The following inequation must hold true:

$k \cdot v_u \neq -1$ for $U_1 \neq U_2$.

NON-INTRINSICALLY SAFE COMPARATOR WITH FAIL-SAFE MONITORING

As a further design concept of an analog comparator, a configuration as shown by fig. 7 will be discussed. It contains a non-intrinsically safe comparator available as an integrated circuit (IC) and a fail-safe monitoring circuit.

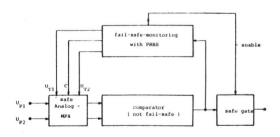

legend:
PRBS = Pseudo-Random-Binary-Sequence

MPX = Multiplexer

U_{T1}, U_{T2} = test voltages

U_{P1}, U_{P2} = voltages from the process

Fig. 7. Not fail-safe comparator with fail-safe monitoring

According to this concept the non-intrinsically safe comparator is checked by a fail-safe monitoring circuit with a PRBS-device controlling a safe analog multiplexer. The operating mode of the configuration can be described by feeding the input signals U_{P1} and U_{P2} - caused by the process - to a safe analog multiplexer, just as the test signals U_{T1} and U_{T2} which are generated by the PRBS-section. The output signals of the analog multiplexer are then processed by a non-intrinsically safe comparator; its output signal is then fed to the fail-safe monitoring device so that it can be decided whether the operation of the comparator is functionally right or not. The output of the safe monitoring device enables a safe gate for the output signal of the comparator device. The analysis relevant to safety of this configuration shows, however, that the responsibility for safety is shifted onto the analog-multiplexer, the PRBS-monitoring device as well as the safe gate. The checking procedure as described may by repeated so fast that an output relais - used as a safe gate - does not drop. But then one has to proof the safety of the analog multiplexer, the PRBS-device, and the safe gate. The analysis relevant to safety shows that a safe realization of the analog multiplexer, the PRBS-section, and the safe gate presents considerable difficulties.

ANALOG COMPARATOR WITH DIGITALIZATION OF THE INPUT VARIABLES

The configuration to be discussed is represented in fig. 8 and corresponds to the concept according to fig. 3. It can be seen that the two voltages to be compared, U_1 and U_2, are parallelly fed to analog-digital-converters, so that the comparison of analogous signals has been transfered to a comparison of digitalized information.

Assuming a mere 8-bit-analog-digital converter, a sophisticated digital comparator is needed for the digitized input voltages. Special difficulties arise when proving the safety of the whole configuration, because extensive electronic circuits and special procedures - according to a given tolerance zone - will be needed.

Although it would be possible to use a microprocessor to realize a given tolerance zone, one will succeed in developing suitable hardware circuits operating with a special binary code. Thus we can avoid the well-known difficulties to get a safety proof for software. The analysis relevant to safety shows that there is responsibility for safety not only to the A-/D-converter, but also to the digital comparator.

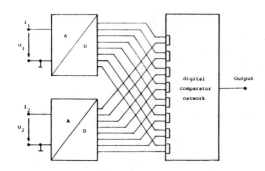

Fig. 8. Comparison of analogous voltages by means of digitalization

PROPOSAL FOR A FAIL-SAFE ANALOG COMPARATOR

The analysis relevant to safety leads to a configuration which is based on a patent application according to /1/. It can be used in the field of safety technique, such as in nuclear power plants as well as in traffic systems. The configuration is characterized by the fact that, by means of a two-channelled comparator configuration, the comparison of analogous voltages is reduced to a digital comparison. The responsibility to safety is passed over to the digital comparator, and one is able to use a fail-safe relais as a digital comparator, if there are no severe limitations with respect to time. This configuration is shwon in fig. 9.

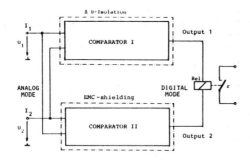

legend: EMC = electromagnetic compatibility
Rel = fail-safe relais
ΔΘ = temperature shift

Fig. 9. Double-channelled fail-safe comparator for analogous signals according to the diversity principle

Furthermore the configuration according to fig. 9 is characterized by the fact that both comparator channels were differently designed. That means that there is used the design principle of 'd i v e r s i t y ' in connection with hardware circuits relevant to safety. There are different measures of diversity when designing a multi-channelled system:

- (1) low-level diversity by using different constructional elements, but following the same circuit diagram; that means: using the same physical principle to realize a required function

or

- (2) high-level diversity by using different physical principles to realize a required function. (One will not succeed in finding in any case different physical principles to solve a problem in multi-channelled systems, but sometimes it is possible).

The presented configuration was designed according to item no. 2.

Drift phenomena, as can occur under various environmental conditions, are mastered by adapting each of the two comparator channels in a different way to the environment:

- one comparator is equipped with special thermal isolation w i t h o u t any EMC-shielding while
- the other comparator is equipped with special EMC-shielding w i t h o u t any thermal isolation.

Thus common mode distortions as well as drift phenomena will be mastered. Further configurations are presented in fig. 10 without and with potential separation.

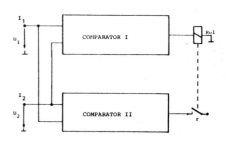

Fig. 10a. Connecting the comparators 1 and 2 by a fail-safe relais Rel

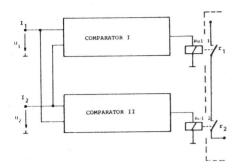

Fig. 10b. Connecting the comparators 1 and 2 by two fail-safe relais in order to guarantee potential separation

SUMMARY

In the field of safety technique the safe comparison of analogous signals is a general problem. This applies to the field of nuclear power plants as well as to traffic systems.

Special problems were pointed out at the beginning of the paper; then the requirements to be fulfilled by a safe analog comparator were mentioned.

Finally different configurations of analog comparators were critically discussed. Here a difference had to be made particularly between realizations without secondary energy and those with secondary energy. As a result of the whole analysis of various concepts of analog comparators, a configuration of an analog comparator was presented which is based on the fact that the comparison of analogous signals is transformed to a comparison of digital signals by a fail-safe relais.

Furthermore the presented configution is characterized by the design principle of 'd i v e r s i t y '. Special investigation should be done in the near future to quantify the advantage of diversity as an important design principle in the field of safety technique.

REFERENCES

/1/ Schildt,G.H., Gayen, J.T. Patent application, em 104/78
/2/ Kraus. Patent application em 74/78
/3/ Schuck, Rieger, Kirchlechner Patent application, em 74/78
/4/ Fricke, H., Schildt, G.H. Conception of Safety and Realization Principles, ATRA-Conference, Indianapolis, USA, April 1978
/5/ Jentsch, D. Grundsätzliches zur Sicherheitstechnik, Präsentation of 'LOGISAFE', AEG-Telefunken, Oct. 1977
/6/ Fricke, H., Schildt, G.H., et al. Sicherheit und Zuverlässigkeit von Nahtransportsystemen, edited by SNV, Hamburg, Dec.1978
/7/ - VDE 876 (EMC), Geräte zur Messung von Funkstörungen, part 1, Sept. 1978

DISCUSSION

Six: Mr. Schildt, you use a double system with two comparators which are not fail-safe. If the signals you compare, are equal, both channels will derive a '1' signal. It may be, that in one channel a failure occurs "stuck to 1". How can you detect this failure?

Schildt: With a certain probability we are able to master this event by using the principle of diversity.

Ehrenberger: Are we able to sum up your presentation as follows? You have two channels, neither of the channels is fail-safe, but they are diverse. If single failure occurs even if double or more failure occur you have a certain chance that this failures are detected due to the diversity, but you don't exactly know the probability.

Schildt: Yes, I would say so.

FAILURE DETECTION IN MICROCOMPUTER SYSTEMS[1]

B. Courtois

*Institut d'Informatique et de Mathématiques Appliquées de Grenoble,
BP 53, 38041 Grenoble Cedex, France*

Abstract. This paper deals with the study of the efficiency of simple mechanisms, in order to detect failures in a microcomputer system : watchdog, memory violation and invalid operation codes detectors. Although detection cannot be immediate, the detection time may be short, but without diagnosis concerning the location of the failure. The necessity of the derivation of failures hypothesis, not only from either physical or electrical schemes, but also from the physical implementation of LSI circuits, is illustrated, and a model for the distribution of the detection time associated to the most likely failures is derived.

Keywords. Safety. On-line testing. Microcomputer.

INTRODUCTION

This paper deals with the study of the efficiency of simple mechanisms dedicated to the detection of hardware failures in microcomputer systems. These detectors are a watchdog periodically executed by the software, a mechanism detecting the memory violations and another one detecting the tentatives to execute invalid operation codes. These mechanisms are prefered to massive redundancy : duplexed or TMR systems may be used, for example in order to increase the safety (1), but when applied to LSI circuits, it is necessary to take care of the possible non-increase of the reliability functions due to the unreliability of the added circuits. If, for example, voting circuits were available in 40 pins packages, we would need at least 5 supplementary LSI circuits for voting in order to implement a TMR structure (and 15 for triplicated voters) on 40 lines. In the past, there was no major problem, even with non-LSI voters, because of the number of circuits needed to implement a CPU. But now the reliability benefit is not evident and other methods have be studied.

The major disadvantage of the proposed mechanisms is that detection is not immediate. But many applications, and many more in the future (let us take for example all the domestic apparatus which use or will use microprocessing) can support this fact, of detection time is confined in reasonable limits. And this disadvantage is counterbalanced by the hardware overhead, which is reduced to a minimum : the implementation of a watch-dog does not need a large amount of hardware. In the same way, a mechanism detecting invalid accesses to the memory, need only very small Hardware, from the existing circuits for normal decoding of adresses (depending on the used circuits, such a mechanism need sometimes only the wiring of pins together). It is quite easy to implement the detection of invalid op-codes in the instruction decoder, and to force the CPU in a STOP-state (the aim of the studies being also useful for the design of new LSI's).

In order to measure the efficiency of the detection mechanisms, we used a simulation system specially devoted to the insertion of hardware faults. More precisely, all experiments were made with the MOTOROLA 6800. The best simulation would have been one made at the lowest level. But such a simulation would be quite inaccurate for time reasons (and costs ones...) Therefore, a functional simulation has been made, taking into account some parts of the internal structure. In particular, the 59 invalid op-codes of the 6800 have been studied (2). This is important because of the non-previsible behaviour of the circuit (the official handbook of the Manufacturer said). For example, although somme op-codes are equivalent to existing instructions, some others have a more sophisticated effect : the instruction code 87 stores the content of acccumulator A at the address (Program counter + 2) ; B3 is equivalent to accumulator A + memory complement → accumulator A and three codes lead the CPU to read sequentially the memory (the only way to stop it is the restart : interrupts have no effect).

---

[1] This work was supported by CNET (Centre National d'Etudes des Télécommunications), LANNION, FRANCE.

The set of units that may be affected by failures include busses and registers. The model of failure is the stuck-at one, and for each unit any bit may be s. a. 0 (1) from time t1 to time t2. This point will be discussed in a next section.

RESULTS OF EXPERIMENTS

Experiments have been made with an operational real-time program, several hundred thousands instructions having been simulated. Yet 150 single (logical) solid failures have been studied, each of them being simulated 1 to 30 times and results as follows.

For all failures affecting the program sequencing, there is always detection by at least one of the three detectors (in most cases : two or threee). Detection times range from times corresponding to some instructions, to times corresponding to some hundreds instructions, depending on the length of the watch-dog. It is to be noticed that 100 instructions represent about 400 microseconds (resp. 200 microseconds) with a 1 MHz 6800 (resp. 2 MHz 6800). More precisely, the behaviour of programs have been studied.

In most cases, the detection is very fast if it is made by means of memory violation or invalid op-codes mechanisms. In other cases, the failure is detected by the watch-dog, the expected value of detection time being equal to the half length of the watch-dog. This is due to the fact that, very quickly, the program enters a loop. This loop may be either existing in the program (and the program never exists because of failure) or artificially created by the failure (a JUMP for example is created from another instruction). Usually,
. in case of memory violation, the necessary is shorter than the time necessary to reach the beginning of the loop,
. the loop exists in the program more often than it is created,
. the number of instructions executed in the loop is relatively small : some tens up to a hundred,
. the memory area concerned with this loop may be made of parts of program functionally independent : some   bytes of an instruction routine followed by bytes belonging to another routine. It is thus understandable that detection by invalid op-codes or memory violation is done in a short time, since as soon as the program enters a loop, the detection by these mechanisms is no longer possible.

The invalide op-codes detection mechanism has been particularly sutied. The following parameters have a great influence on the detection time : the coding of invalid op-codes with respect to the coding of other invalid op-codes, the coding of invalid op-codes with respect to valid op-codes (the dynamic frequency of valid op-codes), the coding of valid op-codes with respect to other valid op-codes.

These results developed in (3) are useful for the design of instructions sets. They have been used for the design of the MOM 400 (a monolithic 4 bit microcomputer (4) and the P 68 - a super 6800 (5) -which are two microprocessors designed in the Computer Architecture Group of Imag Laboratory).

For failures affecting only the operative part, there is no detection (except in special cases when for example an accumulator register is used as a loop index). But if this operative part is used for programs sequencing, detection times become equivalent to those obtained for failures affecting the sequencing. Some experiments were made with the index register of the 6800 system. If this register is used only to address data, there is no detection. But if the program uses some JSR (jump to subroutine) with the indexing addressing mode, we get good detection times.

DISCUSSION ABOUT FAILURE HYPOTHESIS

Up to now, the results that we obtained do not allow us to conclude that the proposed mechanisms would detect all real failures that may appear in the chip. But they encourage to refine the failure hypothesis that were made in order to determine a more realistic set of faults for simulations. Another discussion has to be initiated about the respective probabilities for different failures (it could be possible to differentiate the quality of detection depending of the appearing probabilities).These two points will be illustrated by examples and a model of the distribution time for (at least) the most likely failures will be briefly presented.

1. Refining the failure hypothesis

It is well known now that the stuck-at model is not sufficient in order to take into account realistic failure hypothesis (failures resulting in bridging for example). A first step is to consider electrical schemes instead fo logical schemes. In (6) it is shown that failures represented on the electrical scheme are not present on the logical one, and vice-versa. But the physical implementation of circuits is also to be considered, as the following example will illustrate Among many other failure mechanisms in N-MOS technology, a defect may affect a contact between aluminium of the top level and diffusion of the low level. The corresponding failure mode is an open circuit. Figure 1 represents two memory calls of the ACCU A register of the 6800 (depletion type). The important detail is that the contact aluminium-diffusion of 2 load MOS for the 2 corresponding writing MOS is common for the 2 correspondings bits. This is important, because a signle physical failure affecting this contact will result in a multiple logical failure. The multiple failures are, and they only are :

bit 1 and bit 2 stuck-at "1"
bit 3 and bit 4 stuck-at "1"
bit 5 and bit 6 stuck-at "1"
if we restrict the hypothesis to a single physical failure (which seems to be a natural hypothesis).

2. The most likely failures

When looking at the literature, it appears that the most likely failures affect the fragile chain of a chip : logic for input-output, bond at die, Al or Au wire, bond at post. The percentages are varying : for example from 33 % in (7) to 89 % during tests in (8). Data reported by a French system manufacturer are 78 % for Intel CPU's. Figure 2 represents an example of such a failure : the defect affects the push-pull output transistor of the read/write control line of 6800. It is difficult to precise whether the short is direct between the + 5 volts aluminium area for bonding, through the diffusion, or between the + 5 aluminium and the silicium gate. But the failure made is a stuck-at "1" of the read/write pin, i.e. the CPU is always reading. Depending on electrical characteristics (existence of pull-up's, precharge of busses, ...) a break affecting a bond at die or post, or a wire, will also result in a stuck-at pin in most cases.

For all these failures, each time the CPU will try to execute an instruction, four cases are possible :
a) an invalid op-code is encountered, or will be presented to the decoding ROM,
b) the executed op-code is such that the memory protection mechanism, detects the failure,
c) the executed op-code is such that a loop is initiated (and neither the invalid op-code mechanism nor the memory protection will detect the failure),
d) all other cases.

Considering that probabilities of cases a, b, c, d, are constant (for one failure), an approximate model (approximate because of the non real linearity between time and number of instructions) may be built which allows the derivation of some variables.

The detection time is very much linked to the values of the parameters, i.e. for example the coding of the valid and invalid instructions. Figures 3 & 4 represent the distribution of the detection time, according to the following parameters :

probability of case a) : p_1
b) : p_2
c) : p_3

p_1	p_2	p_3	FIG
0.05	0.01	0.005	Figure 3
0.05	0.01	0.02	Figure 4

On y-axis are probabilities and on x-axis are detection times ; the maximum value is equal to 840 cycles. If p_3 is low enough, the distribution is approximately concentrated between 0 and 280 cycles. For the value of 840 cycles, we can for example state that :
. for $p_1 + p_2$ = 0.01 or 0.02; detection time is distributed through the interval 0 - 840 cycles,
. for $p_1 + p_2$ = 0.1
  - if p_3 = 0.005, detection time is concentrated beween 0 and 190 cycles,
  - if p_3 = 0.01, time is concentrated between 0 and 230 cycles;
  - if p_3 > 0.01, time is distributed through the entire inerval,
. for $p_1 + p_2$ = 0;15 and $p_1 + p_2$ = 0.20, with p_3 < 0.02, the detection time will be less than 140 cycles.

An important question arises about the adequacy of such a model. A lot of experiments have been made to compare the empirical results to the calculations given by the model. An example is the following :
From experiments, we get for one failure (20 times) (detection time by number of instructions) :
. mean value : 21.9 - standard deviation : 6 when failure is detected by invalid op-code mechanism
. mean value : 43.5 - standard deviation : 27.4 when failure is detected by the watch-dog
. mean-value : 29.3 - standard deviation:20 when we take into account the two mechanisms (no detection by memory protection mechanism).

Two equations of the model give :
p_1 = 0.0354
p_3 = 0.0026
an with these values for P_1 and p_3, other equations give for example :
general mean value of detection time : 28 instead of 29.3
probability of detection by invalid op-code : 0.82 instead of 0.65
probability of detection by the watch-dog : 0.18 instead of 0.35

This failure is detected by the invalid op-code mechanism, when a load ACCU instruction is executed. This instruction is often used in programs. That is the reason of the good detection by the invalid op-code mechanism.

Let us give another example :
experiments → mean value when detection by memory protection : 32.9
mean value when detection by the wath-dog : 106.8
Two equations of the model give :
p_2 = 0.0034
p_3 = 0.0186

With these values for p_2 and p_3, other equations of the model give : general mean value of detection time : 98.9 instead of 95.5
probability of detection by memory protection 0.12 instead of 0.15
probability of detection by the watch-dog : 0.88 instead of 0.85

The results given by such a model are fairly well in accordance with the experiments. From a static determination of parameters (without simulation) the aim of such a model will be to predict the detection time associated with some failures (at least the most likely to occur).
In the first example given above, the static frequency of the load ACCU B instruction is about 0.02, while a dynamic frequency would be higher. This is to be compared to the value 0.0354, although the frequency to be used would be a dynamic one in presence of failures.

CONCLUSION

The work reported here is obviously not finished, but the results obtained encourage developments in several directions. One of them is the adequacy of the failure hypothesis, in the sense that we briefly illustrated. Another direction is the development of an "art of programming" in order to use the operative part for the sequencing of programs (but if some circuits are note used in such a manner, they have to be tested periodically, in a conventional manner, for example when the watch-dog is executed). A critique concerns the execution points of the watch-dog : programs have to be modified in order to insert the execution points. Studies have to be initiated on this point, in order to get an automatic insertion.
Finally, let us notice that the proposed type of detectors will be experimented on the CANOPUS project (9). This project is carried out jointly by the IMAG Computer Architecture Group and the CNET Architecture and Command Group, and is concerned with the design of an entirely distributed electronic switching system. This ESS will include, in a final version, some hundreds of "self-detecting" microprocessors, allowing to study the efficiency of detection/signaling mechanisms, through experiments using failed chips, or chips failed by "micro-surgery".

REFERENCES

(1) Courtois B. (1976) Safety and reliability of non-repairable systems, Digital Processes, 2, 4.
(2) Nemour M., Codes-operation invalides de la famille 6800, Note technique, Grenoble (1979).
(3) Courtois B.,(1979) Some results about the efficiency of simple mechanisms for the detection of microcomputer mal functions, Proc. of the 9th IEEE Fault-Tolerant Computing Symposium, Madison, June 1979.
(4) Guyot A. M. Vergnault M., (1978) Spécifications du MOM 400, EFCIS, Grenoble
(5) Presson E., (1978). Spécifications du P 68, Thomson/CSF/SESCOSEM, Grenoble.
(6) Crouzet Y.,Galiay J., Rousseau P., Vergnault M., (1978). Improvements of the testability of LSI circuits, European Solid-State Circuits Conference, Amsterdam.
(7) Glaser and Subak Saarpe, Integrated circuits engineering, Addison Wesley Publishing Co.
(8) Fitch W., (1976). Extended temperature cycling of plastic and ceramic IC's with thermal shock preconditionning, Proc. of Reliability Physics.
(9) Anceau F., Courtois B., Lecerf C., Marinescu M., Pons J.F., (1979). Project of a distributed architecture for the command of an electronic switching system, Proc. of the International Switching Symposium, Paris, May 1979.

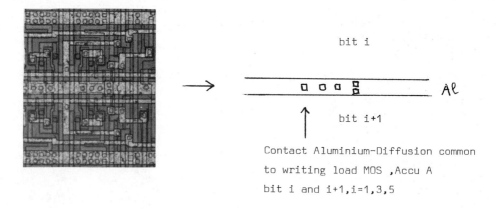

Contact Aluminium-Diffusion common to writing load MOS ,Accu A bit i and i+1,i=1,3,5

Figure 1

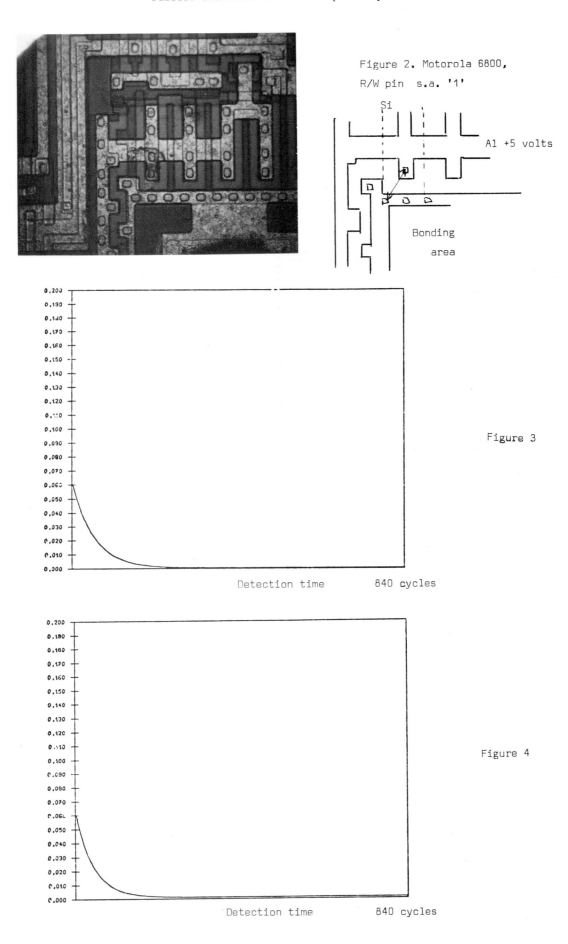

Figure 2. Motorola 6800, R/W pin s.a. '1'

Figure 3

Figure 4

DISCUSSION

Ehrenberger: One particular difficulty with failures of microprocessors has to do with the possibility of short circuits to the substrate. Did you consider this? Did you find figures about the probability for the occurence of these faults? Did you get figures on the probability to detect there faults by your supervision method?

Courtois: There is also a possibility of interaction between different levels. Yet we are able only to simulate logical faults. We have now to try, it's not so easy, to simulate other faults, starting from a low level on the physical defects and trying to insert these faults at logical level, because it is not possible to simulate at the physical level.

Schüller: We have made an investigation about the possible faults of very simple gates with two inputs and one output like NAND-gate or NOR-gate and we found out that - depending on the internal structure - almost all 16 possible two-input (failure) functions can be effects of a single failure. These failure effects have different probabilities of occurence. In fact at least five or six possibilities are not neglegible. So I suppose if you make failure assumptions for higher integrated circuits you must have in mind that nearly every functional effect is possible i.e. any input-output transformation is possible.

Courtois: That's true. But not always. For example, for the memory cells of accumulator, it is very simple to obtain all the failure modes for that parts. It's quite easy for regular arrays, so for all the registers of 6800. For the ROM for the decoding of instructions it is quite easy also. But for all the random parts which are not regular may be it is more difficult.

Ryland: It seems, that the safety related computer system must do two things: it must virtually never perform an incorrect action, but it seems equally important that our computer must perform a correct answer as well. Even if a detected failure exists you have no means of ensuring a correct answer. In other words, if you need the redundancy anyway to ensure the correct answer, than perhaps there is no point in going to special failure detection methods to avoid redundancy in finding a fault.

Courtois: It's not necessary to have redundancy for the detection of failures.

Ryland: No, I said that we need the redundancy anyway. Then perhaps there is no advantage in going to special checking methods.

Schüller: This is a question of failure detection time. If you have no special failure detection methods, - depending on your process - a failure may remain undetected for a long time.
If the failure is in one unit and shows its failure effect only after a long time then the probability increases that a second failure of the same effect occurs in the second unit and thus your system may fail.

For this reason it is necessary to have special failure detecting methods and not only redundancy.

Courtois: I think that redundancy may be used not only for the detection but also for the implementation of gracefully degrading systems.

TEST POLICY VS MAINTENANCE POLICY AND SYSTEM RELIABILITY

L. F. Pau

E.N.S. Télécommunications, 46 rue Barrault, F 75634 Paris Cédex 13, France

Abstract. This paper is concerned with the specification to be given to an automatic test system (ATS) (test frequency distribution, ATS reliability, false alarm rate), in order to interact with the monitored system's reliability and the maintenance policy. A decision rule is introduced to select between unscheduled repair or overhaul in order to minimize the maintenance costs per unit operational time. This decision rule is derived from a Markov model. Results have been applied to the case of an airborne equipment, and to a computer controlled telecommunications equipment.

Keywords: Maintenance policy, ATS, Reliability, Markov model, Monitoring, Calibration intervals.

INTRODUCTION

Independently of the technique used for failure diagnosis or detection, it is essential to investigate the relations existing between the operational availability of a system, the maintenance decisions, the automatic test system diagnostic performances, and the system reliability. The testing is assumed to be a random checkout policy of known probability distribution, it is assumed to be imperfect. In addition to a preventive maintenance to be carried out after a duration T, repair as well as overhaul may be carried out in case of unscheduled maintenance. A simple repair does not extend the potential life, whereas an overhaul is supposed to do it.

The issue adressed in this paper, is to study the effect of the ATS performances on a decision rule among the alternatives H_0 (repair) and H_1 (overhaul) in case of a catastrophic failure detected by the ATS, as to minimize the expected maintenance costs per unit of operational availability.

This decision rule is studied in terms of the parametric influence of:

- the life length T
- equipment and ATS reliabilities
- distribution of the overhaul or repair time
- distribution of the storage times
- probability of incorrect diagnosis and test
- average unit prices, a per repair and A per overhaul.

OPERATING AND MAINTENANCE PROCEDURES

The system is assumed to operate continuously until one of the following cicumstances happens:

A. the system undergoes a checkout (without removal), which is performed with an automatic test system (ATS) (Ref. 5); the check-outs are randomly distributed over time at a rate μ_{01} (t)

B. or: the system is improperly pulled-down, for accessability considerations to other equipments; it is sometimes reinstalled without testing.

If no failure is detected by the ATS, either after checkout or removal, the system is put back into operational status; if it had not been removed, this is done immediately; if not, the system transists through an on-site inventory.

If a failure has been detected by the ATS (correctly or improperly), the system may either undergo repair (on-site), or overhaul (off-site). After repair or overhaul the system is stored into one of two possible inventories before being returned into service. The equipment condition regenerates only when the preventive maintenance also is perfect.

STATE PROBABILITIES

At any time T, the system can be found in one of the following eight states:

E_0: System in service and in operational status.

E_1: System under test without removal.

E_2: System withdrawn from service because of replacement.

E_3: System pulled down.

E_4: On-site repair after a negative test outcome.

E_5: Off-site repair after a negative test outcome.

E_6: Off-site storage of systems in operational status.

E_7: On-site storage of systems in operational status.

Let $p_i(t)$ be the probability for the system to be in state E_i at time t, $i = 0, 1, \ldots, 7$.

TRANSITION RATES

Define the following transition probabilities:

$\mu_{01}(t) \Delta t$ = Prob $(E_0 \to E_1$ during $[t, t + \Delta t] | E_0$ at $t)$ represents the rate at which a system in service is being connected to the test equipment, without being pulled down.

$\lambda(t) \Delta t$ = Prob $(E_0 \to \dot{E}_0$ during $[t, t + \Delta t] | E_0$ at $t)$ represents the time-dependent failure rate of the equipment.

$\mu_{14} \Delta t$ = Prob $(E_1 \to E_4$ during $[t, t + \Delta t] | E_1$ at $t)$ represents the rate at which a system under test, eligible for maintenance due to the outcome of the test, can be repaired on-site.

$\mu_{15} \Delta t$ = Prob $(E_1 \to E_5$ during $[t, t + \Delta t] | E_1$ at $t)$ represents the rate at which a system under test, eligible for maintenance due to the outcome of the test, can be repaired off-site.

$\mu_6 \Delta t$ = Prob $(E_6 \to E_0$ during $[t, t + \Delta t] | E_6$ at $t)$ represents the rate at which an operational system stored off-site is being put back into service.

$\mu_{70} \Delta t$ = Prob $(E_7 \to E_0$ during $[t, t + \Delta t] | E_7$ at $t)$ represents the rate at which an operational system stored on-site is being put back into service.

$\mu_{03} \Delta t$ = Prob $(E_0 \to E_3$ during $[t, t + \Delta t] | E_0$ at $t)$ represents the rate at which a system in service is being pulled-down on-site.

$\mu_{02}(t) \Delta t$ = Prob $(E_0 \to E_2$ during $[t, t + \Delta t] | E_0$ at $t)$ represents the rate at which a system is being removed from service for replacement or because of aging, μ_{02} is time dependent because of aging.

EXOGENEOUS NON CONDITIONAL PROBABILITIES

Due to reliability considerations, or in connection with the existing maintenance organization, the following probabilities are defined:

$F(t)$: system reliability, $F(t) = \exp(-\lambda(t) \cdot t)$.

$G(t)$: test equipment reliability.

r_1 : probability of detecting the system failure with an operating test equipment.

r_2 : probability of assigning a faulty system to off-site repair.

r_3 : probability of putting back into service a system which was just pulled-down, without carrying out any test.

$\gamma(t)$: probability that the test system delivers a negative output, for one of the following reasons:
- the test equipment itself has failed;
- if not, the system is operational but the test equipment detected a failure (false alarm);
- or: the system had failed, but the test equipment did not detect the failure.

$$\gamma(t) = 1 - G(t) + G(t)(1 - r_1) F(t) + G(t) r_1 (1 - F(t))$$

STATE EVOLUTIONS

It can be shown that the evolutions of the state probabilities $p_i(t)$ are governed by the system of equations:

$$\frac{dp_0}{dt} = -(\mu_{01}(t) + \mu_{02}(t) + \mu_{03}) p_0(t) + G(t)(r_1(1 - \lambda(t)) + (1 - r_1) \lambda(t)) p_1(t) + \mu_{60} p_6(t) + \mu_{70} p_7(t)$$

$$\frac{dp_1}{dt} = -(+ G(t)(r_1(1 - \lambda(t)) + (1 - r_1) \lambda(t)) + (1 - r_2) \gamma(t) \mu_{14} + r_2 \gamma(t) \mu_{15}) p_1(t) + \mu_{01}(t) p_0(t) + (1 - r_3 \mu_{03}) p_3(t)$$

$$\frac{dp_2}{dt} = \mu_{02}(t) p_0(t)$$

$$\frac{dp_3}{dt} = - p_3(t) + \mu_{03} p_0(t)$$

$$\frac{dp_4}{dt} = - p_4(t) + (1 - r_2) \gamma(t) \mu_{14} p_1(t)$$

$$\frac{dp_5}{dt} = - p_5(t) + r_2 \gamma(t) \mu_{15} p_1(t)$$

$$\frac{dp_6}{dt} = -\mu_{60} p_6(t) + p_5(t)$$

$$\frac{dp_7}{dt} = -\mu_{70} p_7(t) + r_3 \mu_{03} p_3(t) + p_4(t)$$

INITIAL CONDITIONS AND STEADY-STATE SOLUTION

If the system is operational at t = 0:

$p_0(0) = 1 \quad p_i(0) = 0 \quad i = 1, \ldots, 7$

It is noted that the differential equation above does not fulfil the ergodicity requirement in all cases:

$$\lim_{t \to +\infty} \frac{dp_i}{dt} = 0 \quad i = 0, \ldots, 7 \text{ independently of } p_i(0)$$

This is essentially due to the imperfections of the test equipment, even if $\mu_{02} = 0$.

Ergodicity, as well as singularity conditions are of secundary importance here, because, we are only interested in finite horizon numerical solutions; this finite horizon is equal to the preventive maintenance period T.

MEAN TIME BETWEEN UNSCHEDULED REPAIRS

In case of preventive maintenance actions of period T, the MTBF has to be replaced by the MTBUR.

a) No test and removals

$$\text{MTBUR}(T) = \int_0^T R(t)\,dt$$

Example: Exponential process:
$$\text{MTBUR}(T) = (1 - \exp(-\lambda T))/\lambda$$

Example: Weibull process:
numerical results in (Ref. 10).

b) Test and removals

$$\text{MTBUR}(T) = \int_0^T p_0(t)\,dt$$

where $p_0(t)$ is computed as solution of the system of differential equations in section 5.

CRITERIA APPLYING TO PREVENTIVE MAINTENANCE DECISIONS

Either of the following maintenance decision rules can be considered:

a) Overhaul or replacement of the system if it is found to be in a state of deterioration exceeding a control limit;

b) minimize the cost per service time unit, when the cost of replacement or overhaul and the costs of being in the various degraded states are taken into account (incl. inspection costs);

c) maximize the time between replacements or overhauls subject to the assumption that the reliability does not exceed some upper limit;

d) minimize the life cycle costs per item, including procurement, testing, maintenance, inspection and operations.

COST MINIMIZATION OF THE DECISION BETWEEN UNSCHEDULED ON-THE-SITE REPAIR AND OVERHAUL

Consider:

t: time elapsed since the last scheduled or unscheduled overhaul

A: mean unit overhaul cost (off-site)

a: mean unit repair cost (on-site)

T: preventive maintenance period

a) General case

Assume that the test equipment actually detects a failure at time t; the decision is between:

H_0 : on-site repair at cost a, if $t \leq \theta$

H_1 : off-site overhaul at cost A, if $\theta < t < T$

The maintenance strategy proposed here, is to select the decision threshold θ, which minimizes the mean maintenance costs per unit time of operational life between unscheduled maintenance. In case of on-site repair, the costs can only be spread out over MTBUR (T), because the authorized residual life remains unchanged; in case of overhaul, the costs should be spread out over MTBUR (t + T), because of the extension in the authorized residual life.

$$\theta : \text{Min}_t \left(\frac{a}{\text{MTBUR}(t)} p_4(t) + \frac{A}{\text{MTBUR}(t+T)} p_5(t) \right)$$

The left hand-side criterion may be computed numerically for various values of t, and a limiting curve produced for the optimum θ values in the (T, a/A) coordinate system.

The numerical solution, obtained by forward integration, will assume $\mu_{01}(t)$, $\lambda(t)$, $\mu_{02}(t)$ to be specified exogeneous functions; these functions may very well be discontinuous, in order to allow for practical reliability or maintenance requirements.

Examples:

1) $\mu_{01}(t)$, test rate, is specified equal to zero over subintervals of [0, T], corresponding to the unavailability of an on-site ATS

2) $\mu_{02}(t)$, removal rate, is specified equal to zero over subintervals of [0, T], for example during the contractual life of the equipment.

Such conditions may lead to the existence of local minima during the optimization of θ, due to nonmonotonous variations of $p_4(t)$ and $p_5(t)$. Practical considerations will usually lead to the selection of the largest locally optimum solution θ.

b) Exponential_and_Weibull_F-distributions without_testing

Full numerical results and charts for θ were obtained in (Ref. 10).

FAILURE DETECTION AND TESTING

This procedure assumes randomly distributed in time ATS activation, and negligible testing time. It also assumes essentially in-service on-line testing, carried out by either redundant hardware of software (watchdog). In case of error no contamination to other processes is assumed.

An optimal test policy may also be derived, by expressing the transition probability matrix $Q(t, u)$ (see section 5) for the process, as a linear combination of two transition matrices $Q_0(t)$ and $Q_1(t)$. $Q_0(t)$ will have a bias towards the breakdown state, and $Q_1(t)$ towards the normal state, while u is a control variable. If $u = 0$ the UUT is not tested, and it is operated according to $Q_0(t)$ with random testing.

If $u = 1$, the UUT operates according to $Q_1(t)$, and is being maintained-

$$Q(t, u) = (1 - u) Q_0(t) + Q_1(t)$$

The problem is to find the optimal control u^* based on the number of failure detections by the ATS, as to maximize a criterion over $[0, T]$.

APPLICATION TO ATS DESIGN SPECIFICATION

The problem of specifying the reliability $G(t)$, probability of false alarm $(1 - r_1)$, and the probability of non-detection of an automatic test system is essential for the overall operational readiness of many aerospace and telecommunication systems, which operate as described in Section 1. Moreover, these imperfections have a crucial effect on the preventive maintenance policy.

The results of section 5 have been used to select $\mu_{01}(t)$, $G(t)$ and r_1 in agreement with specifications on the operational availability represented via the MTBUR (T) or via the plot of $p_0(t)$. It is essential, in case of airborne equipments, to find out whether adding an ATS with fairly poor $G(t)$ and r_1 figures will not in fact reduce the operational availability, as compared to the random maintenance policy. Another concern is not to maximize $\mu_{01}(t)$, which is the rate at which the monitored system is being tested, because of side-effects related to the ATS reliability.

APPLICATION TO THE SPECIFICATION OF RECALIBRATION INTERVALS

The issue is here to detect failures which correspond to measurements out of specifications, because of drift, etc. .... For a given maintenance policy, one has therefore to optimize the rate $\mu_{01}(t)$ at which recalibration is carried out.

REFERENCES

[1] AGARD, Integrity in electronic flight control systems, AGARDograph No. 224, AGARD - AG - 224, Paris, April 1977

[2] R. Barlow, F. Proschan, Mathematical theory of reliability, Wiley, NY, 1967, pp. 84 - 118

[3] D. Chaudhuri, K.C. Sahu, Preventive maintenance interval for optimal reliability of deteriorating systems. IEEE Trans. Reliability, Vol R - 26, No. 5, December 1977, 371 - 372

[4] I.B. Gertsbakh, Models of preventive maintenance, North Holland Publ. Co., Amsterdam, 1977

[5] IEE, New developments in automatic testing, Proc. Int. Conf. 30. - 2. dec. 1977, Brighton, Publ. by IEE, London, 1977

[6] V.B. Iversen, On the accuracy of measurements of time intervals and traffic intensities, 8th Int. Teletraffic Congress, Melbourne, November 1976

[7] A. Kaufmann, D. Grouchko, R. Cruon, Modelès mathématiques pur l étude de la fiabilité des systèmes, Masson, Paris, 1975

[8] H. Moskowitz, R.K. Fink, A Bayesian algorithm incorporating inspector errors for quality control and auditing, in: M.F. Neuts (Ed.), Algorithmic methods in probability, Studies in the management sciences, Vol 7, North Holland, Amsterdam, 1977

[9] NASA, Systems reliability issues for future aircraft, NASA - CP - 003, 18 - 20 august 1975

[10] L.F. Pau, Révision générale et remise en état (Overhaul and repair), T.R. E.N.S. Télécommunications, Paris, 1970

[11] L.F. Pau, Diagnostic des pannes dans les systèmes (Failure diagnosis in systems), Cepadues - Editions, Toulouse, 1975

[12] C.P. Tsokos, I.N. Shimi (Ed.), The theory and applications of reliability, with emphasis on Bayesian and non-parametric methods, Academic Press, NY, 1977

[13] E.B. Dynkin, A.A. Yushkevich, Markov processes: Theorems and problems, Plenum Press, NY, 1969

[14] F. Liguori, Automatic test equipment: hardware, software and management, IEEE Press, NY, 1974

[15] L.F. Pau, Failure diagnosis and performance monitoring, Marcel Dekker Inc., NY, 1979

[16] A. SEGALL, Optimal control of noisy finite-state Markov-processes, IEEE Trans., Vol AC - 22, No. 2, April 1977, 179 - 186

[17] C. Bellon, Etude de la dégradation progressive dans les systèms répartis, PhD thesis, Institut National Polytechnique de Grenoble, 14. sept. 1977

DISCUSSION

Aitken: I would like to ask Professor Pau on what basis he calculated the MTBF to 6000 hours. This is a lower figure than we would anticipate and we would expect microprocessors to have a better MTBF much more as have many computers.

Pau: The estimated MTBF depends clearly on the design and the environment which you consider. It depends also on the quality control procedure which you apply, e.g. whether you have software errors and burn in testing or not. Especially for aerospace applications the MTBF depends on whether you have special hardware and similar types of processes.

However, the figure of 6000 hours just considered is only a given value which could be reasonable. You may be aware of the evaluation carried out by the Intel-Corporation for the militarized version of its microprocessors which indicates even lower figures for the MTBF in some cases. The Rome Air development Center (R.A.D.C.) has carried out an effort in order better to determine what reasonable values may be expected; they are often not much higher than 9000 hours, according to what I heard. This is simply to stress again that one has to be careful when using such microprocessors in 'stand alone' operational applications, especially in severe environments.

$\gamma(t) \overset{\Delta}{=} 1 - G(t) + G(t)(1-r_1)F(t) + G(t)r_1(1-F(t))$

Figure 1. Graph of the state transitions

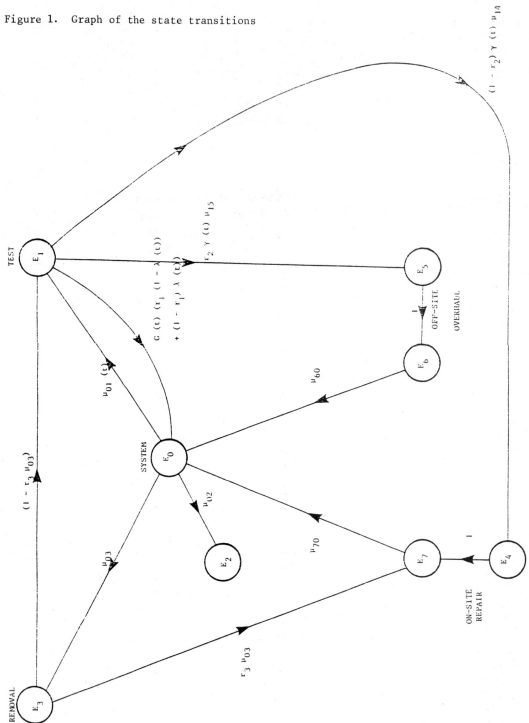

OPTIMISATION OF A SERVOSYSTEM

T. Adjarov

Department of Electrical Engineering, Institute for Computing Technics, Sofia, Bulgaria

Abstract. The optimization of thecapstan-servosystem in the tape drive in a data entry device can be achieved by modeling and investigating the system by digital methods. The present paper deals whit a model, creating by dividing the system into small indivisible units a computer program by the means and methods of Continuous System Modeling Program. The way of preliminary system testing and optimizating is pointed as the way for avoiding loses and unreliable performance in the practice.

Keywords. Servomechanisms; system analysis; modeling; system simulation; optimisation.

INTRODUCTION

The safe and reliable performance of a system largely depends on the methods by which it has been designed. The better and the more completely it has been tested and verified before being introduced in industry, the more reliable and safer its further performance in practice. Hence, while the system has not yet assumed a final shape, its optimisation and testing is of great importance so that some weak points, unnoticed during its design can be elliminated while it is in blueprint.

A method of such an investigation and optimization of an electromechanical system is presented in this paper on the basis of a tape drive in a data entry device. Furthermore, the methods tested for modeling and optimizing given as an example of such a comparatively small system, can also be applied to larger and more complicated systems, automatic control systems in industrial processes and others. The degree of dividing such a system into minimally dividable units with a minimal number of changeable parameters is important in this case since it is essential to obtain a clear notion of their action upon the system as a whole or a separate block of it. In other words, we must have an overal idea about the realities in the system.

COMPOSING OF THE MODEL

In making up the model of this paper, some of the techniques given by Diakov and co-workers (1976) have been used. The simulation of the process occuring in the system under modeling also results in economy of labor, materials and means.

This manner of work enables the designer of systems requiring safe and reliable performance to make his solutions in accordance with the latest demands both of the market and technology, preserving at the same time the reliability and the safety of the device under different conditions.

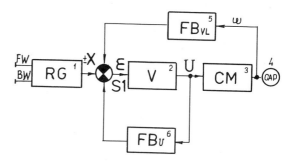

Fig. 1. Functional structure of the system

The functional structure of the system consist of a ramp generator (RG), amplifier (V), capstan motor (CM), voltage and velocity feed backs (FBu) and (FBv).

The ramp generator gives two reference voltages establishing the direction and the velocity value. The obtained control deviation, resulting from the addition of this reference voltage and the feedback voltages determines the velocity of the electromotor by the amplifier so that a constant velocity can be maintained during performance. The control deviation also regulates the start-stop times. The system described is thus rather generalized and does not allow evaluating the influences of the parameters determining the action of the units described, since each unit is a rough generalization of the function of a given block. Therefore a more detailed investigation of the system is necessary as well as its division into units with a minimal number of acting parameters so that their subsequent optimization can be undertaken and their influence on the performance of the system can be taken into account.

DETAILED MODELS AND SIMULATION RESULTS

A detailed model of the ramp generator is shown on Fig. 2.

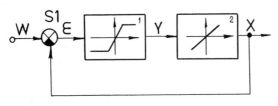

Fig. 2. Ramp generator structure.

The configuration of this system is similar to the common system of an on-off controller, as it is given by Fieger (1971). In this particular case such a controller is unsuitable because of the rough regulation of the output variable X. This lack of precision will have a further adverse effect. For this reason the on-off controller is replaced by a limiter and the controlled system by a precise integrator as it was described by Adjarov, Topalov (1974).

With the help of the means and methods of the IBM developed CSMP, the ramp generator was modeled and investigated through a computer. The adequacy of the built up an computer modeled and investigated system to the real system has been proved. The print-plotted results are given in appendix 1.

The basic signals acting in the ramp generator are given there as an on-off controller at the beginning and as a first order lag at the end of the process.

The optimization of the process is carried out by artificially placing a disturbance Z in the system. The disturbance could practically appear at the output of the system. The disturbance Z is placed as follows in the formation of ε.

$$\varepsilon = KL \cdot [W - (X + Z)] \qquad (1)$$

When the disturbance is given, the simulation of the processes takes place for different values of the constant KL, defining the slope of the linear part of the limiter. In the table given below we can see the start-stop times, the response times of the given disturbance for the three values of KL, and the percentage deviation. These values have been computerized.

TABLE 1 Simulation Results

KL	50	100	200
Tst/ms	30	25	23
Tr /ms	4,5	2,5	1,5
%	0,028	0,014	0,01

From the results generalized in TABLE 1 it is evident that the deviation itself is insignificant compared with the disturbance chosen to be of such a value that it can even change the sign of ε and Y. As is seen the increase of KL has a favourable effect on the dynamics of the system A further increase of KL leds to the limiter assuming a switch function, which results both in instability of the system and in a decrease of the noise imunity.

Simulation without changing the field of the linear action have been made, i.e. only the value of the limiter has been changed

TABLE 2 Simulation Results without Changing

KL	50	100	200
Tst/ms	50	25	12
Tr /ms	-	2	1
%	-	0,014	0,013

Here Tr and the percentage deviation for KL=50 cannot be seen very well since the values of X in the time for constant velocity change all the times,i.e. the instability of the control process itself is rather high even without an external disturbing influence.

The first kind of simulation appears to be more appropriate than the second one as they have been given by the integrator time constant. The practical change of KL for the first kind of simulation is easier to realise.

The structural scheme given in details on Fig. 3 can give an idea about the conections between the units of the amplifier and the electromotor.

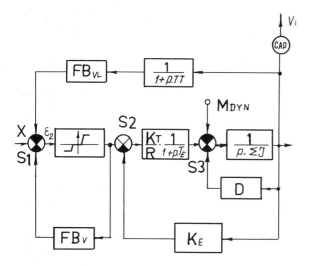

Fig. 3.Detailed structural scheme

The signal ε_2 is the result of the algebraic sum of the reference voltage,multiplied by a coefficient in order to obtain the necessary velocity and the stabilizing feed back voltages.This control deviation reflects the real state of the system at any time.However a power amplifier is necessary,which can allow reversing the velocity of the motor.For practical reasons,such a non-linear characteristic is needed which can eliminate certain small disturbing voltages,resulting from offset and bias instabilities of the summing preamplifier.

The modeling of the electromotor is based on three relationships:

1. The voltage in the electromotor ;

2. The conection between the current, passing through the anchor winding and the torque;

3. The velocity and the acceleration in rotating on the one hand,and the moments of inertia and the load torques on the other.

The last relationship is of a great importance for this case.The rotary motion is defined by Newton's second law as follows

$$\frac{d^2\theta}{dt^2} = \frac{\Sigma M}{\Sigma J} \qquad (2)$$

Using the above relationships and adding the influence of the friction torque through the constant D,given by the motorproducer,we can obtain the complete model of the motor,similar to the given one in Electrocraft's handbook (1973).

The summing of torques takes place in S3.Here we should bear in mind counteraction which the electromotor encounters in overcoming of the tension arm spring,acting in the capstan radius.Besides this,the action of torque caused by the friction of the tape on the block magnetic heads is taken into account in the model.Early investigations (Luarsabishvili,1969) showed,that this torque depends on the nature of the material of the magnetic heads.

In the first place ,the sum of inertia moments includes the one of the motor anchor,and to a smaller degree, the moment of the capstan.The large and changing inertia moment of the reel is not included here because of buffering by the tension arms.The action of the arms inertia moments on the capstan servo is insignificant as this could be proved by using the formulas given by Michnewitch (1971). The whole system is fed back by a dc tachogenerator coupled to the motor and to the servosystem through a T-shaped filter.

The thus composed model was prepared for computer processing my means of the CSMP and several simulation runs were carried out for establishing its adequacy to the real system.After the stage of specifying and fixing parameters and configuration was covered, we set about changing certain parameters and even replacing a given block from Fig. 1 by another.The motor in the initial model was replaced by another one,which was a practical requirement.A part of the current and start-stop-reverse velocity print-plots are given for comparison in appendices 2 and 3.Print-plots A are of the velocity and the anchor current of the initial motor respectively. The forcible change of this motor for reasons of economy worsened the performance of the system and the reliable action of the whole device. This process was theoreticaly simulated by means of the described model and the results are expressed by the print-plots B.The current during the transient processes is of a slightly ramp form which results in a prolonged start and a slow approach of the velocity to the final value.This is both unoptimal and unstable.The instability of the system and the maintained velocity is significant compared to the velocity of print-plots A, this being one of the major factors for the decrease in the reliability of the data recording.To improve the performance of the system simulation runs were carried out for greater and deeper feed back.The feed back giving good results approaching the initial ones and easy to realise in production was selected (print-plots C). An improvment of the transient processes was observed and the instability of the velocity maintained during performance was decreased.

CONCLUSIONS

The methods and techniques applied here for an alectromechanical system can also be applied to similar systems.The application of the mentioned princips is also possible in other fields of the technology.In this way we obtain the basis for preliminary testing and foreseeing the system performance under given conditions , and this could eliminate undesirable surprises resulting in loses and unreliable performance.We are enabled to respond flexibly to the increasing demands of modern technology by limiting undesirable and even failure situations.

REFERENCES

Adjarov, T.,T. Topalov (1974). About some problems in the ramp generator in a data entry device.Proceedings of IIT, VI A ,23-32.
Diakov, D.,T.Adjarov, B.Raitschev, I.Sarafov, N.Rainov (1976). Investigation of the stable and reliable performance of tape drives.Proceedings of the VIth National Conference for Computer Technology,Varna.
Electrocraft Corp. (1973).An Engineering Handbook,Hopkins,Minesota.
Fieger, K. (1971).Regelungstechnik, Grundlagen und Geräte,H & B Meß- und Regeltechnik.
Luarsabishvili, D.G. (1969).Friction performance of the magnetic tapes and the elements of the tape drives.Technic & Television,9,27-31.
Michnewitsh, A.W. (1971).Tape Drives, Energia,Moscow.

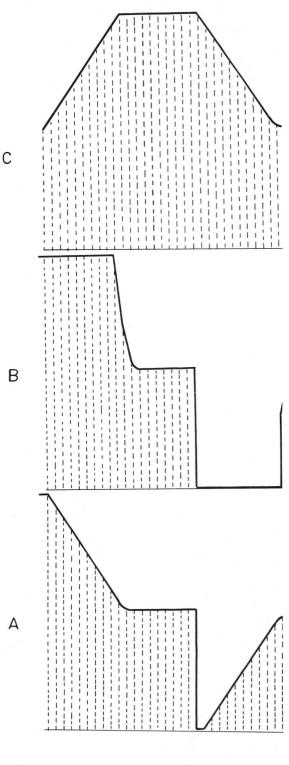

App.1

Smith, I.C. I would like to ask if you have some figures for time to carry out the digital simulation compare to the time to carry out the same run on the original analog system.

Adjarov, T. The time for the whole simulation run was about two or three minutes. The ratio between the digital simulation time and the real time was about 500 till 700 to one.

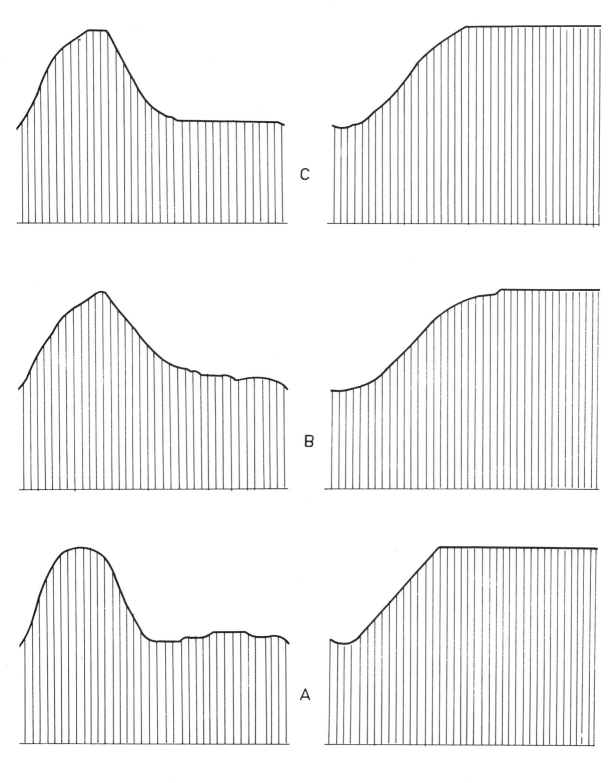

App. 2   App. 3

INSPECTION OF PROCESS COMPUTERS FOR NUCLEAR POWER PLANTS

G. Glöe

Technischer Überwachungs-Verein Norddeutschland e.V., Hamburg, Federal Republic of Germany

Abstract. Process control computers are used in nuclear power plants for very different tasks. Depending on the significance for safety the demands on faultless construction and reliability are more or less urgent. Those demands are the main topic of the paper.

While analysing process control computer systems in nuclear power plants you will find, with respect to the usual separation between hardware and software, some more problems in the field of software than in the hardware. The reason will be discussed.

Developments in computer technology may change this situation a little. There are strong efforts to simplify software production, and as a result it may become more easy to check software. Due to increasing integration and hectic changes it will be more and more difficult to obtain full range information about hardware, to process these informations and to verify them with reasonable expense and time.

Keywords. Power station control; nuclear plants; computer testing; computer debugging; programming control; safety; licensing; standards.

INTRODUCTION

Up to 1975 the numerous control-problems important to safety in nuclear power plants were solved by conventional, that means, not programmable electric and electronic systems.

Since then conventional control logic equipment is displaced to a greater extent by process control computers.

Installation and operation of those safety related control systems require the approval by the licensing authority. In consequence, an expertise is generated by order of this department.

The main points of this discourse are experiences and difficulties in the production of those expertises about process control computers on one hand and on the other hand in realization of the corresponding tests, which are used to get knowledge of the system features.

The following parts enclose three different fields, in which process control computers are used, and the requirements they have to meet. Finally the tests of the hardware, the software and the complete system will be described.

The three application fields

- reactor protection system
- control rod control system
- trouble logging

are altogether safety related, but are considerably different in the grade of danger in case of malfunction.

If an equipment is classified as "safety related" it has not to fulfill automatically a catalogue of fixed design rules.

According to the idea of a "concept of graduated damage-precaution" the demands on a system are measured at the extent of possible damage in case of system failure.

Consequently, the reactor protection system has to satisfy the hardest requisitions. It works in the closed loop mode and must at least contain three redundant subsystems.

The control rod control system works closed loop, too, but without redundant subsystems.

Finally the trouble logging equipment will be discussed. It works on line open loop. Here a supposed failure does not mean immediate danger.

HARDWARE - SOFTWARE

Before going into the details of the above mentioned applications, it will be discussed, why the more significant problems are to be found in the field of software.

These difficulties are depending on the following subjects:

- For the conventional hardware exists an admitted test-mode, which is suitable to check at least the critical components. For achieving comparable, generally accepted procedures for the software, till now there was too little need of programs of high reliability and too little knowledge about appropriate test methods.

- The available guidelines for reactor protection systems offer some criteria to form an opinion on the hardware, but do hardly help in analyzing the software. They more deal with the control of random failures mainly by redundancy in space etc., than with the control of systematic faults.

- Generally hardware is produced in greater quantities. Therefore the analysis can rely on experiences from other applications. The safety related software with its extreme demands on reliability is throughout a special purpose construction. Comparable configurations for other uses do not exist.

Developments in computer technology may change the problems in hardware and software a little. There are strong efforts to simplify software production, and as a result it may become more easy to check software. Due to increasing integration and hectic developments it will be more and more difficult to obtain full range information about hardware, to process these informations and to verify them with reasonable expense and time.

The importance of faultless function of hard- and software is exceeded by the weight of the integral system characteristics and of the cooperation between the two parts. In particular the exact timing is difficult to be proved, furthermore the self-checking features induce an unusual tight gearing between hardware and software and thereby prevent a completely separated consideration.

COMPUTERIZED REACTOR PROTECTION SYSTEMS

In reactor protection systems some physical variables are permanently recorded and logically processed into actions against dangerous situations.

Both, unwanted actions and suppressed actions as a result of malfunction may cause danger. Back-up systems are not available.

About reactor protection systems exists a lot of partly very detailed rules and guidelines (Fichtner, 1977).

For our work in particular the following german regulations are important:

"Sicherheitskriterien für Kernkraftwerke" (Safety criteria for nuclear power plants) (BMI, 1977).

"RSK-Leitlinien" (Guidelines of the reactor-safety-commission) (RSK, 1979).

"KTA-Regel 3501" (Rule 3501 of the nuclear technology board) (KTA, 1977).

Because of the good general view and the accentuation of the fundamental aspects the "Sicherheitskriterien für Kernkraftwerke" offer a good entry into the subject.

The matter is particularized especially in the "KTA-Regel 3501".

The demands on construction by the rules have caused characteristic features in the reactor protection system. The most evident ones are:

- Redundancy, to preserve the function in case of a single failure,

- Separation in space between the redundancies to avoid damage of the whole system e.g. in case of a fire,

- Decoupling of the redundancies against each other and against systems which do not belong to the reactor protection, to prevent the transfer of disturbances,

- Self checking equipments, to discover failures at once,

- Fail safe reaction against faults in the reactor protection to ensure the reactor against indirect and direct endangering,

- High availability of the reactor protection system to be always master of troubles.

The licensing procedure and the tests of a reactor protection system is worked up step by step in the same way as in a rod control- and in a trouble logging system. By doing that the inquiries become more and more detailed.

The usual sequence till the commissioning of a system contains expertises about:

- concept
- erection

and certificates, based on

- suitability test of components
- approval of documents

- functional tests
- test of commissioning and operation.

In special cases the sequence may vary.

We have used a number of quite different test procedures for the examination of reactor protection computers and for finding out the qualities of systems.

The development to practicability and the scientific foundation of the actual test methods were done mainly by the "OECD Halden Reactor Project" in Norway and by the "Gesellschaft für Reaktorsicherheit" in Munich.

Mainly the following tests were realized:

- analysis and corresponding tests of the user programs (Ehrenberger, 1973; Goldstine 1963)

- statistical tests with a hybrid-computer (Ehrenberger, 1972)

- tests concerning especially the time behaviour by help of an analogue computer (Roggenbauer, 1975)

- tests of the self-checking software by hardware error simulations

- reliability examinations.

The selection of the test procedures accorded with the need to find a group of tests, which figures as many computer characteristics as possible. The little number of methods, whose course and assertion were sufficiently examined, raised limitations.

By no means the application of these tests is prescribed for other systems. As long as the aspired characteristics can be shown, the test method is not very relevant. In the inspection procedure, too, the result is more important than the method.

In the work with process control computers (Glöe, 1977) some aspects recently have got particular importance.

- According to the great capacity of computers it is possible to increase the quality of data processing.

- The transfer of duties from several conventional electronic branches into a single computer means a concentration of risk and as a result extraordinary high requirements at the reliability of the equipment.

- There is no software-redundancy without software-diversity. However, diversity means considerably increased expense of work in development, inspection and maintenance of a system.

- Computers normally are sensible against short voltage failures. The consequences are high demands on power supply and emergency power supply.

- The exact determination of the time to produce an action is uncomparably more difficult than in conventional systems.

CONTROL ROD CONTROL COMPUTERS

Rod control computers are single computer systems working on line closed loop as a part of the control assembly control.

Failures in rod control computers may cause an inadmissible power distribution or intolerable insertion of reactivity. The reactor protection system can be considered as a back-up system for this computer.

However, the reactor protection limits possible damages on a higher level than the rod control computer does.

For the rod control system there is only one rule available, the "Sicherheitskriterien für Kernkraftwerke" (BMI, 1977). Because this guideline does not go into the details, a great margin is offered for the selection of the concept of this system.

The actual rod control computers are fairly similar to reactor protection computers in its hardware and in important parts in the software, too.

This similarity made it possible to reach back to the numerous examinations and experiences with reactor protection computers. Thus, for the most part the test procedures were well known and only in special cases suitable methods had to be searched and tested in their practicability for the judgement of the rod control computer.

The positive judgement of a single computer for the safety related task in the rod control system could be found on behalf of the following facts:

- the rod control computer has not to be pernamently available

- the separation between computer and process is a fail safe reaction

- with great likelihood an approved self-checking software in case of faults separates the computer from the process.

I think that in future, too, for tasks in rod control systems or comparable applications, single computers can be judged positively. An assumption is the proof, that the self-

checking programs will notice all faults. But
this extensive proof can only be justified
if there are very high requirements or if
numerous similar systems are planned which
are based on the same hardware. These efforts
normally are too extensive if hardware
modifications frequently will cause changes in
self-checking software.

In this case other computer systems may be
used. The safety of these configurations will
not be provided by a well performed and
completely checked software package, but will
be assured by sufficient redundancy. A
division of these systems may be accomplished
according to the degree of redundancy and
diversity of hardware and software. By
increasing redundancy or diversity there will
be a decreasing requirement for tests. For
checking a system in this manner we need
reliability data about hardware and software.

With the increasing number of computers a lot
of those data should be available before long.
So we think, it will be possible to proof the
qualification of computers by a statistical
approach, e.g. for use in control-assembly
control.

The expense for examination and tests of the
first rod control computer in the TÜV
amounted 1.5 to 2 years for one man.

For the 2nd system, which is tested now, the
expense will be 6 months to 1 year.

TROUBLE LOGGING

After technical trouble informations are
needed of the cause and the stress imposed on
the material. Thus, the logging of incident
proceedings is safety related. Nevertheless
a defect in the logging devices means no
direct danger.

As in control rod control for trouble logging
too, there are only very few german rules or
standards with rather general requests
(BMI, 1977; RSK, 1979). Computers are n't
mentioned at all.

Dealing with an expertise about trouble logging
computers we presume, that the system should
encounter the following perhaps self evedent
requests:

- For trouble logging single purpose computers
  should be used. In this way other programs
  cannot spoil trouble logging procedures.

- The computers must be capable to process all
  changes of variables needed for incident
  review. That are some thousands, mainly
  binary variables, in certain plants some
  hundreds of them may change within a few
  milliseconds.

- The time resolution of the system must
  always allow the separation between the
  cause and the consequences of a disturbance.
  That means a time resolution of some ten
  milliseconds.

- At least two redundant computers should be
  used. The argument for this request is not
  the need to encounter the single failure
  criteria but the need to get a sufficient
  availability.

- Besides the output on a line printer all
  informations should be stored on a tape or
  disk. So you can accomplish more
  sophisticated data evaluations automatically.

- There is no need for self-checking features.

- The trouble logging devices should not be
  affected by plant disturbances.

- Principles of reliable programming should
  be used (Ramamoorthy, 1974; Vogels, 1975).

As already mentioned in the introduction a
malfunction of the trouble logging does not
mean immediate danger. Therefore the checks
of trouble logging may be rather modest,
compared e.g. with reactor protection system.
Especially, we think there is no need for the
formal and complete proof of program
correctness.

For trouble logging it seems to be sufficient
to demonstrate correspondence of the above
described demands with system features by

- check of data sheets and specifications

- execution of some functional tests which
  usually are part of the scope of supply.

Some closing remarks will be given now upon
documentation.

The positive result of a qualification
procedure for a computer system depends on
the quality of the computer just as well as
on the quality of the documentation.
Probably some excellent systems can not be
accepted because the documentation is
insufficient. Till today documentation is a
very critical problem especially in software
development. The designer of hardware
equipment is forced from the beginning to
give tolerable documentation, otherwise the
prototype will look quite different than it
should do. The programmer very often works
as a solist and thinks about documentation
as an annoying thing.

Producers of computers and programs very
often are convinced of the high quality of
their product by a lot of positive
experiences. But for an expertise we are
forced to prove our opinion by exact

documents. Therefore the pure potencial of experiences of the producer or the expert may be enough to get a good impression. Indeed, experiences must be documented, too, moreover there must be a possibility to ensure the real contents of these documents. A significant number of such documents, e.g. about MTBF, MTTR or the frequency of program errors, undoubtedly would solve a lot of problems in checking process control computers for the nuclear licensing procedure.

BMI (1977). Der Bundesminister des Innern (1977). Sicherheitskriterien für Kernkraftwerke Bundesanzeiger, 206, 1-3.

Ehrenberger, W. and E. Soklic (1972) A Hybrid Method for Testing Process Computer Performance. Enlarged HGP Meeting, Loen.

Ehrenberger, W. (1973). Zur Theorie der Analyse von Prozeßrechnerprogrammen. Laboratorium für Reaktorregelung und Anlagensicherung, Garching.

Fichtner, N.; K. Becker and M. Bashin (1977). Cataloque and classification of technical safety standards, rules and regulations for nuclear power reactors and nuclear fuel cycle facilities. Commission of the European Communities, Brussels - Luxembourg.

Glöe, G. and H. Eggert (1977). Zum Einsatz von Prozeßrechnern für sicherheitstechnisch bedeutende Aufgaben in Kernkraftwerken. TÜ, 18, 192-196.

Goldstine, H.H. and J. v. Neumann (1963). Planning and Coding Problems for an Electronic Computer Instrument Part 2, Vol. 1 - 3, John von Neumann collected works, Vol. 5. New York. pp 80 - 235

KTA (1977). Reaktorschutzsystem und Überwachung von Sicherheitseinrichtungen. Sicherheitstechnische Regel des KTA, KTA 3501, Fassung 3/77. Carl Heymanns Verlag, Köln.

Ramamoorthy, C.V. and others (1974). Reliability and Integrity of Large Computer Programs. In G. Goos (Ed.), Lecture Notes in Computer Science, Vol. 12. Springer Verlag, Berlin-Heidelberg-New York.

Roggenbauer, H.; J. Ueberall, TH. Korpas and U. Scot Jørgensen (1975). Testing a Computerized Protection System with an Analogue Model. Enlarged HPG Meeting, Sandefjord.

RSK (1979). RSK-Leitlinien für Druckwasserreaktoren, 2. Ausgabe, 24. Januar 79 Gesellschaft für Reaktorsicherheit, Köln.

Voges, U. and W. Ehrenberger (1975). Vorschläge zu Programmierrichtlinien für ein Reaktorschutzrechnersystem. Gesellschaft für Kernforschung, Karlsruhe.

DISCUSSION

Genser: What is about your opinion to use standardized software like it is also in hardware, if you want some approval of some authority?

Glöe: I think in software for nuclear power plants we have not standards at all, so we cannot use standardized software. Moreover there is not any request to use only standardized software. We have the rules which are mentioned in my paper and if the software is in accordance with these rules that is enough.

Genser: In history of hardware it was common that you have some test equipments and by gaining experience during the past time you found some standardized way and if you obey the experience of the past then you get very easy the approval of the authority but not for very new design. In this case you get a certificat for a limited time for testing your hardware perhaps in some testing field but not for going operation in reality.

Glöe: In Germany there is the possibility to go into operation with rather new hardware systems if they succed in a suitability test. After having succeeded in this test a new hardware component can be used in the unclear power plant.

Genser: Why do not you believe that it is also possible for software?

Glöe: I mean it could be possible for software too, but I do not know actual test methods for software like they are generally accepted for hardware.

Genser: How is the qualification of people involved in the system considered by your certification method?

Glöe: I do not bother about the qualification of people. I am only concerned with the qualification of computers.

Lauber: Up to now, we were discussing on safety of computers. But I think considering the special application as you do, you also have to talk about security. How do you approve security of computers? By security I mean to prevent that human interaction from outside may disturb these programs which should be safe.

Glöe: Everything is done that a saboteur from outside cannot enter the plant, but if anybody inside the plant wants to sabotage, we think you actually do not have any real chance to prevent that.

Lauber: How many computers are installed and in operation up to now of the three kinds you mentioned?

Glöe: In Germany there are two reactor protection computer systems working online but open-loop, to get experiences, that means six computers in two systems. Rod control computers are working online closed-loop in four plants. Trouble logging computers are installed in each plant.

Ehrenberger: I first refer to the contribution of Dr. Genser. Certainly man machine cooperation and the role of the operator is very important. There are special conferences on that subject. To deal with this is the task of TC5 of the European Computer Workshop. This question is not investigated in TC7. In the reactor business it is assumed that the operators have their appropriate qualification, before they are allowed to use the computers.
Secondly I want to say something to Professor Lauber's contribution. The question of security has been investigated very thoroughly in connection with nuclear power plants. The results of the investigations, however, is being kept secret for obvious reasons.

Last not least some remarks to Mr. Smith's contribution? Obviously the amount of information to be processed by the operator in case of disturbances is enormous. At GRS-Garching therefore a computerized disturbance analysis system has been developed. It consists of a computer that contains a model of the plant and that receives all the signals that are displayed in the control room. If any deviation from the normal state occurs, this system is able to find out by menas of its model
a) what the ultimate cause of the disturbance was
b) what the possible consequences of the disturbance are, in particular what the most probable disturbance propagation is
c) what the operator reasonably should do in order to overcome the abnormal situation
d) what the possible consequences of intended operator actions could be.

This system is currently in the state of final testing. It shall be installed at the PWR Grafenrheinfeld.

Parr: I heard recently a description of a safe way of interpreting commands which could be made available to an operator. Do you foresee the possibility of a general purpose interpreter type language perhaps on a standard machine which could be given a standard approval because this control system is always in store. Although control are left to a high level program the interpreter shows the operations are all safe. The interpreter is well used, well tested, well documented. The control programs which can be written by perhaps engineering staff rather than computer scientists are always safe. They may not be correct, but if they are incorrect they lead to safe actions. Is that a possibility?

Glöe: It would be sufficient for us, but for the operating company the availability of the plant is important. Therefore the programs also must be correct. That is not concerned with safety, but the software must be correct to be a helpful instrument to generate electricity.

Maki: You say that a single computer do rod control, if you could prove that the self-checking programs can detect all faults. Do you have some reasons that this would possible.

Glöe: I said in my paper that with great likelihood an approved self-checking software in case of faults would separate the computer from the process. The safety is guaranteed by a back-up system and the control rod control. The reactor protection system is a back-up for the rod control computer.

Smith: In typical fault conditions large amounts of random data arrive very quickly. Do you have guidelines for a filtering system so that the available data storage system conquers this situation?

Glöe: Half a year ago we had the first order to work about trouble logging. Till today there are not any rules because there was no need for them.

AUTHOR INDEX

Adjarov, T. 207
Aitken, A. 11
Albert, T. 161
Bologna, S. 103, 129
Clark, D. W. 153
Cooper, M. J. 153
Courtois, B. 195
Dahll, G. 89
Daniels, B. K. 11
De Agostino, E. 103
Dubuisson, B. 149
Ehrenberger, W. 129
Frey, H. H. 3, 41
Gilbert, M. H. 59
Glöe, G. 213
Gmeiner, L. 75
Grams, T. 65
Holt, H. 139
Konakovsky, R. 81
Lahti, J. 89
Lauber, R. 1
Lavison, P. 149
Levene, A. A. 33

Long, A. B. 161
Mattucci, A. 103
Monaci, P. 103
Mullery, G. P. 51
Nam, C. 161
Pau, L. F. 201
Puhr-Westerheide, P. 117
Putignani, M. G. 103
Quirk, W. J. 59, 153
Reeves, H. 161
Saib, S. 161
Santucci, C. 179
Schäfer, M. 65
Schildt, G. H. 187
Schüller, H. 179
Schwier, W. 169
Smith, I. C. 11
So, H. 161
Straker, E. 161
Taylor, J. R. 95
Trauboth, H. 41
Voges, U. 75, 95